D1348729

C0000020065858

Inheritors of the Earth

CHRIS D. THOMAS

Inheritors of the Earth

How Nature is Thriving in an Age of Extinction

ALLEN LANE
an imprint of
PENGUIN BOOKS

ALLEN LANE

UK | USA | Canada | Ireland | Australia
India | New Zealand | South Africa

Allen Lane is part of the Penguin Random House group of companies
whose addresses can be found at global.penguinrandomhouse.com

First published 2017
001

Copyright © Chris D. Thomas, 2017

The moral right of the author has been asserted

Set in 10.5/14 pt Sabon LT Std
Typeset by Jouve (UK), Milton Keynes
Printed in Great Britain by Clays Ltd, St Ives plc

A CIP catalogue record for this book is available from the British Library

ISBN: 978-0-241-24075-5

www.greenpenguin.co.uk

Penguin Random House is committed to a
sustainable future for our business, our readers
and our planet. This book is made from Forest
Stewardship Council® certified paper.

For
Rose, Alice, Lucy, Jack
and Helen

Contents

CONTENTS

PART I

Opportunity

Prologue
Gains and losses

The Vale of York is a landscape of ecological despair. Billowing clouds emerge as condensation from the cooling towers of distant power stations and then rise above the plain. Fields of wheat, barley and oilseed rape disappear towards its flat horizons. Ploughs and combine-harvesters trundle where brown bears once snuffled through primeval forests and wild cattle wallowed in peaty swamps. It is one of the most intensively farmed parts of England, and England is one of the most densely populated countries in Europe.

Yet when I survey this landscape from my study window, I see a green world punctuated by scarlet specks of poppy flowers swaying in the breeze. Coppery pheasants are strutting in search of fallen seeds, dowdy field voles are scurrying back and forth below tussocks of grass, and camouflaged greenfly are jostling for position as they suck juices from the veins of young leaves. Their enemies are on the prowl too. Rufous foxes are sniffing out a pheasant meal, hovering kestrels are watchful for the slightest movement of a vole, and dangerous harlequin ladybird beetles are laying waste to the greenfly. Lines of helmet-sized molehills hint at the life below ground, where microbes and fungi decompose fallen leaves and velvety moles consume gritty worms. A water vole is burrowing through a mat of pigmyweed in the pond.

The apparently denuded vale is full of species, and many are world travellers. The harlequin ladybirds originated in Asia; the small-leaved pigmyweed in Tasmania; poppies came as contaminants of grain from continental Europe; and kestrels arrived under their own volition, able to catch their rodent prey once the forest was cleared. In contrast, the tunnelling moles have survived generations of tumult as the ancient forest was converted into farmland; now they burrow beneath road verges and grassy lawns. Although these players and the scenery of life's game – the species and habitats – have changed on account of the presence of humans, the basics of biology remain. Regardless of their origins, the plants growing in the fields, roadsides, ditches, hedgerows, gardens and scattered copses still capture energy from the sun and convert it into leaves, rendering the world green; animals consume plants and their seeds and in turn are killed and eaten by other animals. Decomposing plants, animals and faeces are recycled as nutrients, used by next year's growth. The rules of life continue, save *Homo sapiens* is now a key player.

It is the same elsewhere. Familiar creatures have taken the human-modified Earth by storm, be they talkative yellow-billed Indian myna birds that are now at home in Florida, Japan, Sumatra, Madagascar and Australia, agile mice that started life in Asia and then spread throughout the cities, towns and farmsteads of the world, or Australian wattle trees and previously endangered Californian pines that are growing wild in Africa. They are joined by thousands of other mammals, birds and plants, as well as by microbes, fungi, worms, snails, shrimps, insects, fish, toads and lizards. The trickle of successful species[1] taking advantage of human-created opportunities is becoming a torrent. Seemingly, there are as many winners as losers.

We should certainly mourn the losses. The biological and physical processes of the Earth already bear the indelible signature of humanity – this is why scientists are increasingly referring to the present day as the Anthropocene epoch. We have converted a third of the world's vegetation to produce our food, leaving a diminishing space for wild plants and animals. We have altered the great chemical cycles of the Earth beyond their historical bounds, acidified the oceans and changed the climate of the entire planet, threatening any species that cannot adjust. Our ancestors hunted most of the largest land animals to extinction, and we exterminated others by transporting voracious predators and virulent diseases to remote islands. Of those that survive, 13 per cent of bird species, 26 per cent of mammals, 31 per cent of cacti, 33 per cent of reef-forming corals and 42 per cent of amphibians are threatened in some way.[2] A mass extinction is in full swing, and prognoses for the future seem dire. For these reasons, we have gone so far as to describe ourselves as the scourge of the Earth, and as exceeding our planetary boundaries.

On the other hand, we are still surrounded by large numbers of species, many of which appear to be benefitting from our presence. If some species are declining but others are thriving in this human-altered world, is the prognosis really as bad as the doom-laden message of biological decline? It is important to recognize the ways our actions threaten the Earth, of course, and I will highlight many losses that have already taken place and continue to this day. However, in many respects, nature is coping surprisingly well in the human era. We should not ignore the gain side of the great biological equation of life.

It is vital that we take a broad view, considering all the evidence, if we are to draw conclusions that run somewhat counter to the 'paradigm of biological decline' that predominates among ecologists, environmentalists and conservationists. So, in this book, we will embark on a round-the-world tour of the planet's diverse continents and far-flung islands, visiting locations where my research has taken me over the past several decades. (I refer to changes in the oceans only in passing, in the interests of keeping this volume to a manageable length and because it is not my area of expertise.)[3] Our journey will take in tick- and leech-filled tropical forests, bone-cold high mountains, oceanic archipelagos overrun by foreign species and landscapes devoted to the production of our food; each with its own separate human history. I also draw on insights from a wider scientific literature, based on research that has been carried out in a broader range of locations across the Earth. This is important because we live in a globalized world: our greenhouse-gas emissions warm the climate everywhere; we consume food and use timber that may have been grown on the opposite side of our planet; and species are being transported in our wake. We need a truly global perspective to understand the ramifications of our own actions.

We also need to set today's changes in their appropriate historical context, which involves time spans much longer than we are used to thinking about in our everyday lives. This is necessary because the story of life on Earth is one of never-ending change: be that the arrival and disappearance of species from a particular location (ecological change) or the longer-term formation of new species and extinction of others (evolutionary change). So we examine the responses of species and ecosystems to human impacts over periods that range from years to millennia. We will also reflect on the fates of animals and plants as they experienced ice ages on schedules of tens to hundreds of thousands of years. And we will examine the consequences when different types of animals and plants meet up for the first time, as happened several million years ago, when a land connection was made between North and South America, introducing sabre-toothed cats, four-tusked elephants, camels and tapirs from the north to a southern continent inhabited by marsupial lions, porcupines, giant ground sloths and enormous armadillos. The message is clear. Some species don't make it. Yet, it is equally clear that there are always survivors, and they persist and then thrive by moving across the surface of the Earth to places that are hospitable to them, and by evolving new capabilities. They are still doing so today.

In this journey through time and space, we make some surprising discoveries. Visits to coastal Brazil, to the tropical forests of Mexico and Cameroon and to the thoroughly transformed British Isles reveal that new species have colonized landscapes that contain a mixture of arable fields, pastures, hedgerows, plantations, orchards, logged forests, ditches and towns faster than the rate at which species that used to live in the original habitats have disappeared (as long as there are still sufficient remnants of the earlier vegetation to act as refuges for the more sensitive species). The result of these comings and goings has been that diversity has grown in nearly all regions of the world that cover areas as large as, say, Belgium or the state of Vermont. Moreover, our reliance on agriculture means that we no longer need to hunt for food as our ancestors did – hence, the numbers of bison and other large mammals are starting to rise again in North America and Europe, where the cultural transition from killing to saving animals has developed most strongly. Climate change is playing its part too. More species like it hot than cold, and so the overall consequence of a warmer climate is to raise biological diversity in many parts of the world.

This rise in the number of species in each region has been fuelled unwittingly by human action – as we have colonized all six continents, we have carried different species with us, from horses to hippos, pigeons to pythons, lobsters to lionfish. We are acting as a global glue, bringing together previously isolated biological worlds into one 'virtual continent', a New Pangea. In so doing, we are initiating a modern retelling of what has always been key to the story of life on Earth – the successful succeed. This is seen most keenly on remote islands, where introduced rats, cats, dogs, stoats and human apes have ousted tame, flightless and disease-prone birds and walking bats that have been isolated for millions of years. The continental carnivores and diseases have won, as we will discover in New Zealand and elsewhere. However, most new species that arrive do not displace any others, the consequence of which is that these islands now contain far more species than they used to. The same is true of our continents – the diversity of forest trees and shrubs is increasing in the Alps of southern Europe, for example, because many more immigrant species have established new populations than native species have disappeared. This is the norm – immigration usually increases the diversity of the recipient region.

In the longer run, the generation of biological diversity is down to evolution. The biological riches we see around us today exist because more different types of microbes, plants and animals have evolved than have become extinct over the course of the Earth's history. The Earth formed 4.5 billion years ago, single cells came into existence over 3.5 billion years ago, creatures consisting of complex bodies took off in the last 600 million years, and the diversification of life on land has occupied the last 450 million years. On average, biological gains have tended to have the upper hand, and this increasing variety of life seems likely to continue for hundreds of millions of years to come. Yes, periods when the levels of extinction are high – as they are now – represent major setbacks, but in the end they have provided new opportunities for enterprising creatures that have been able to exploit the new conditions. Nature has come back from mass extinctions before and the variety of life has grown again. Could this happen once more? Living through an age of extinction provides us with a unique opportunity to explore whether there is an evolutionary yang to the yin of human impacts.

It seems that there is – a global-scale spate of rapid evolution is in full flow. We will meet butterflies in the USA and in the meadows of Britain and Europe, and Australian crickets now living in the Hawaiian Islands, which are all evolving new ways of surviving in the human-altered world. Just as some species are thriving in modern times, some genetically distinct individuals and genes are becoming increasingly successful. Even more remarkable, new animal and plant species are coming into existence, an evolutionary signature of the Anthropocene epoch. In fact, the formation of new hybrid plant species in Europe and North America would appear to be faster already than the rate at which previously existing plants are becoming extinct. Furthermore, animals and plants which we have transported around the world are acquiring characteristics that make them less and less like their ancestors. Eventually, they will become separate species. The Earth is poised for a massive acceleration in the formation of new species – come back in a million years and we might be looking at several million additional species whose existence can be attributed to the activities of humans.

This broad geographical, historical, ecological and evolutionary context reminds us that the biological world is in constant flux. Dynamic change means that we face an indefinite future of biological gains as well as losses – often with humans the underlying cause. What are we to make of this human-altered world? The default stance of conservation is to keep things as unchanged as possible or, alternatively, to return conditions to what they used to be, or somehow to make the Earth 'more natural' (which can usually be interpreted as meaning 'with reduced human impact'). Not only are these untenable aspirations while the world's human population continues to grow and each of us consumes more resources, but hold-the-line strategies implicitly dismiss as undesirable the continuing biological gains of the human epoch. Where is the logic in this? Attempting to prevent the establishment of 'alien' arrivals and to kill 'impure' new hybrids so as to maintain our ecosystems and species in some idealized state is not possible, nor is it obvious that the past state of the world is objectively preferable to the new state that is coming into existence.

We need a new rationale for the relationship between humanity and the natural world in which humans are regarded as *part* of nature, given that we, too, have evolved and everywhere on Earth has already been altered by us. We have to work *with* natural biological processes, not against them. There is no point in taking on a never-ending fight with the inevitability of eventual failure. The new philosophy opens the door to a more optimistic approach. It permits us to be appreciative of the biological beneficiaries of the human-altered environment, while remaining cognisant of the many human-caused losses. Keeping as many species as possible alive on our global Ark should still be a primary target for our conservation activities, however, because these species and those that evolve from them are the building blocks from which every future ecological system will be constructed. They will fuel future dynamism.

Rather than swim against the tide of ecological and evolutionary change, we should remember that the old was once new. The story of life is one of diversification and renewal – successful genes and species win the game. It is time for the ecological, conservation and environmental movement – of which I am a life-long member – to throw off the shackles of a pessimism-laden, loss-only view of the world. Why should we not aspire to a world where it is as legitimate to facilitate new gains as it is to avoid losses? But before we get to this, we need to consider the success stories that have already unfolded, and how it is that humans might eventually increase the biological diversity of the Earth.

I

Biogenesis

A rutted track snakes inland from the eastern shores of the Caspian Sea. Clouds of dust bring glistening tears to the eyes of the weary traveller. Mile after mile, a khaki world extends into the distance. The Asian steppe is a tough place for Earthlings, the landscape little more than dry grasses, low bushes and crumbling earthen banks.

Cheep emanates from a dry crack in a bank beside the track.

Cheep, cheep. The sound is surprisingly familiar. Then, a small grey, black and russet object emerges from the hazy sky and disappears into the crack with a flurry of feathers and dust. A cacophony of high-pitched squeaking follows; it is feeding time on the steppe. A few seconds later, its chicks fed, out pops a sparrow. No wonder it sounded familiar.

Its relatively large and pallid appearance identify the bird as a Bactrian sparrow. Roadside 'camel-crossing' signs depict animals with two humps, revealing that this is also the land of the Bactrian camel. The sparrows spend the spring and summer raising young in the open landscapes that range from southern Kazakhstan, Turkmenistan, Uzbekistan and eastwards towards the Bactrian plain of Tajikistan and northern Afghanistan. Then they travel southwards to spend the winter months foraging for grass seeds and dropped cereal grain over the plains of Pakistan and northern India, before travelling northwards the following spring to avoid the oven-like Indian summer. It is a natural cycle, a wild bird migrating to avoid the north when it is too cold and the south when it is too hot.

The occasional intrepid Bactrian sparrow is drawn away from this tough landscape in search of easy pickings. They can be spotted stealing grain from local stalls in Kazakhstan's Baikonur market during the summer months, and from the small Indian village markets that dot Rajasthan's landscape in winter. And this may be how it all started. Somewhere in the swathe of land that today encompasses northern India, Pakistan, Iran, Iraq and westwards to Syria, wild sparrows much like the present-day Bactrian sparrows must have started to find it easier to find food by hanging around the villages that sprang up in the fertile valleys where agriculture first developed. The large and tasty grass seeds – progenitors of wheat – that our ancestors cultivated would have been enticing. Rather than glean small, scattered seeds from the steppes and semi-deserts of western and southern Asia, why not head into town and stuff your crop from a large pile of big fat grains? It is easy to imagine that these bloated individuals might have survived better, and then have been able to raise more chicks than their wilder ancestors. And if the villagers were also making new earthen banks with cracks in them – the walls and roofs of their houses – then why not nest in them rather than fly miles away in search of natural crevices? So it must have been that a group of sparrows gradually abandoned their wild existence, maybe ten thousand years ago, and moved into our villages and towns. They have been our companions ever since.

Its destiny sealed, the sparrow spread around the world from its origin in Asia, moving to wherever we built houses, spilt grain, left bags of food unguarded and deliberately threw out our scraps for them.[1] The wild grey-headed sparrows, with streaked orange-brown and black wings and creamy-grey breasts, whose appearance allowed them to merge into the brownscapes of western Asia, entered our homes and became the familiar house sparrow. These new domestic sparrows remain the same biological species as the truly wild birds that still fly out over the Asian steppe in search of insects and seeds. In some respects, they are unchanged. They continue to pursue insects for their young (often in our yards and fields) and seek out grass seeds (which we call cereal grains) for themselves. They are still living in grey-brown places: our towns – it just turned out that the conversion of the world into a land of farms, villages, towns and cities made the whole world just a bit more like their Asian homeland. They got lucky.

The house sparrow conquered Asia and Europe thousands of years ago, eventually coming to inhabit villages and towns from Portugal in the west to China in the east, and from Sri Lanka in the south to Norway in the north. The familiarity of Ancient Egyptians, Greeks, Romans and Indians with them attests to the long period over which sparrows have been integrated into the human world. They became part of human culture: they were sacred to Aphrodite, the Greek goddess of love, and their exuberant and public copulation led the Indian scientist Varāhamihira to admire the sparrow's virility in the sixth century. Others interpreted their behaviour as vulgar. Sparrow hieroglyphs denoted a sense of being bad and small to Ancient Egyptians. St Matthew's gospel related how sparrows were regarded as of such low value that two would be sold for a single Roman assarion coin.

Somewhat more recently, Queen Elizabeth I's 1566 law classified sparrows as vermin and placed a small bounty on each bird in England. In the eighteenth century, the sparrow was the nursery-rhyme villain who killed cock robin with his bow and arrow. Hymn writer Civilla Martin composed 'His Eye is on the Sparrow' in 1905 to explain how, if the Lord's munificence stretches as far as the insignificant sparrow, he surely must care for any human who would sing his praises. Humans have had a love-hate relationship with the house sparrow for at least five thousand years, and probably for ten thousand.

Having completed its European and Asian adventure, the sparrow was then thwarted, bounded by the Atlantic Ocean to the west, and by the Indian Ocean and the Sahara Desert to the south. These distances were too far to fly. If it was to follow the burgeoning world's human population elsewhere, it needed help. Fortunately, it was at hand. In the nineteenth century, New Yorker Eugene Schieffelin embarked on releasing – in North America – every different kind of bird that was mentioned in the works of William Shakespeare. To help achieve his aim, Schieffelin argued that 'English' sparrows, as they had by then become known, would help control insect pests in the New World. So they were brought to North America to get rid of a 'plague' of caterpillars of the snow-white linden moth on no stronger evidence than that sparrows sometimes eat caterpillars. They failed to control the caterpillars and instead mainly consumed dropped grain, human food scraps, and seeds that could be excavated from

The house sparrow has a contrasting grey crown (opposite, top); the Italian sparrow's crown is brown and it has a neater chin (opposite, middle); tree sparrows clearing left-overs off plates in a jungle lodge at Danum, in Borneo (opposite, bottom); and tree sparrow with its dark cheek patch (above), seeking out scraps on a feeding station for proboscis monkeys (below). House sparrows originated in Asia and are now spread throughout the world, associated with humans. Tree sparrows are also found in Asia, but have become the urban and village sparrow of tropical Southeast Asia; another human-associated success story. Italian sparrows did not exist before humans developed agriculture and towns. They originated when house sparrows spread out of Asia and hybridized with Spanish sparrows.

the piles of horse dung that adorned the streets of New York City at the time. They were released in at least a hundred towns and cities across thirty-nine American states and four Canadian provinces between 1851 and the end of the 1880s.[2]

Sparrows were also introduced to Cuba and Argentina in the 1800s and quickly spread through the Americas. They were released in Australia and New Zealand, and brought to South Africa and Zanzibar, from where they colonized most of the other countries that lie to the south of the Sahara. They even made it to remote islands across the Pacific, Indian and Atlantic Oceans, sometimes intentionally, sometimes as stowaways. The house sparrow became a true world traveller. A fairly localized and apparently quite unexceptional russet brown, buff, black and grey bird that lived in the grassy plains of Asia has become one of the Earth's most widespread species, relying on humans to provide it with transport, food and homes. It is one of the most successful species in the world: inheritor number one of the human-altered Earth.

The close association of humans and sparrows continues to the present day and still generates the age-old mix of fondness and irritation. My childhood education in coarse language was provided by my father's guttural curses as he cleared sparrow nests that were blocking our house's down-pipes; a mass of dry grass, feathers, horsehair, paper and baling twine showered me as I steadied the ladder. Dad found sparrows distinctly irritating, and he would have been delighted that there are only a third as many house sparrows in Great Britain today as there were on that summer day of 1967 when we cleared the eaves. But conservationists and concerned citizens are seemingly on the side of my seven-year-old self, who liked having sparrows nesting in the roof. They want to protect sparrows, even though British house sparrows actually stopped declining around about the year 2000; there are still over 10 million of them in Britain, not to mention half a billion in the world.[3] The house sparrow is one of the most widespread animals on our planet and, if we are to be rational about spending limited conservation budgets, one of the species that least needs our help. This concern for house sparrows is doubly odd because conservationists tend not to look kindly on foreign species – and there would be no house sparrows at all in Britain had it not been for the fact that humans created somewhat steppe-like conditions in the Home Counties. But sparrows arrived long enough ago that we have apparently declared them to be a 'native species' that needs protection. Researchers funded by government, conservation organizations and universities have been spending time and money to diagnose the 'problem', which implies that the preferred state of Britain is to have even more sparrows. More precisely, because the justification for the work is the two-thirds decline between the 1970s and the year 2000, the 'correct' number of house sparrows is taken to be roughly how many there were in the 1970s. Picking any specific date is obviously completely arbitrary. Choose a start date of ten thousand years ago, and the 'correct' number of house sparrows in Britain would be zero. Not to be put off, the recent surge in enthusiasm for sparrows has even been incorporated into an updated 1981 version of English law, which aims to protect all wild birds. In the second Elizabethan age, it is illegal to kill or injure sparrows intentionally, or to remove any of their nests that are in active use.

This research has revealed that the late-twentieth-century decline likely stemmed from a reduction in the amount of dropped grain and weed seeds, insufficient insect food in urban areas to provision their chicks, and perhaps also from an absence of suitable crevices for the birds to use as nest sites in modern buildings.[4] Cleanliness leads to fewer sparrows. The solution was obvious: feed the birds. The bird-feeder business is booming and people have even started putting out trays of expensive mealworms (beetle larvae) so that sparrows can supply their insect-deprived chicks with protein-rich meals. Concerned citizens and local councils have begun to install human-made breeding cavities. You can purchase conventional wooden nest boxes, 3D-printed ones, or hollowed-out bricks that provide nesting sites in the walls of your house and three-storey sparrow houses to provide safe havens. The love that humans have for sparrows seemingly has no end.

But the hate continues too. A century ago, the professional writer and lecturer William Dawson, no fan of Eugene Schieffelin, described the introduction of the English sparrow to the United States as *the most deplorable event in the history of American ornithology*.[5] While European researchers and conservationists try to increase the numbers of sparrows, their North American counterparts contemplate the reverse. You can order sparrow-proof nest boxes and 'Deluxe Repeating Sparrow Traps' online. No doubt it is a profitable business selling anti-sparrow devices to concerned citizens wishing to oust unwelcome foreigners, and pro-sparrow feeders and nesting boxes to their neighbours who would welcome them in!

In America, sparrows stand accused of evicting beautiful native bluebirds from their nesting holes, and even of killing them. As the Michigan Bluebird Society put it: '*House Sparrows are an overly aggressive, alien species of bird that prefers similar habitats and nesting locations as bluebirds. The male sparrow is particularly nasty and will often kill not just the young bluebirds but even the adults and eggs too. House Sparrows MUST be controlled in the habitat your nest boxes are placed in to ensure the nesting success of bluebirds.*'[6] This sounds dire, but the prize is great – the protection of bluebirds. With their brilliant cobalt-blue backs and heads, rusty-red chins and chests and downy, cream bellies, bluebirds are stunning creatures indeed. But should we really inflict capital punishment on sparrows for the horrors that they inflict on luckless bluebirds?

If we care about the future of bluebirds, we first need to know whether bluebirds are declining, as many conservationists have claimed, and whether sparrows are the cause. American naturalists have been counting bluebirds and other native species using standardized methods as part of the Audubon Society's Christmas Bird Counts since 1900, and counts of sparrows commenced in 1951. Their records tell a tale that is not consistent with the fear-mongering of the pro-bluebird, anti-sparrow lobby. Oddly, sparrows have declined by some two-thirds since 1951, just as they have in Britain (probably for similar reasons), while bluebirds have increased by nearly a half.[7]

Closer inspection of the Christmas Bird Counts shows that year-to-year changes in the numbers of bluebirds are not affected by the numbers of sparrows, and year-to-year changes in the numbers of sparrows are not affected by the numbers of bluebirds either. The fact that the number of bluebirds increased while that of sparrows decreased (although they remain more numerous than bluebirds) between the 1950s and the 2000s appears to be coincidence, no

different from the chance of getting one head and one tail when toss-
ing a coin twice. This is not surprising. Only a modest proportion of
all evictions of bluebirds from their nest sites, and a very low fraction
of all bluebird deaths, are likely to be attributed to sparrows, given
that most bluebirds live in rural locations, where they benefitted historically from humans clearing the original forest and opening up the
countryside. In recent years, they have also benefitted from the
deployment of bluebird nest boxes, and this may well have contributed to increased numbers in some areas.[8] If we consider the last two
centuries in the round, both these bird species have taken advantage
of living alongside humans.

So why have sparrows been unfairly blamed for the imaginary
decline of bluebirds? Why do we not equally hate all the native North
American animals – including tree swallows, flickers, hawks, chip-
munks, squirrels and raccoons – that also evict or kill bluebirds?
While it is true that sparrows are guilty as charged of killing some
bluebirds, so are our pet cats. Is the reason simply that sparrows are
clad in dowdy browns and greys and bluebirds are clothed in appeal-
ing iridescent reds and blue, or is it because sparrows are, as the
Michigan Bluebird Society put it, 'alien'?

When it comes to conservation debates, it often seems as though
we have set ourselves apart to act as referees and arbiters of how
nature should be – yet our stance lacks consistency. There is no correct state of nature. In Western Europe, we have declared house
sparrows to be native and desirable, even though house sparrows live
in the region only because humans created the towns and farmland
they thrive in. We apparently want to have even more individuals of
a species whose numbers already run into the hundreds of millions
across the world. Why not just be happy with the numbers we have?
They would not exist in North America either without humans –
there they are treated as alien and potentially harmful. The desirable
state of North America is apparently to have fewer sparrows, or none
at all.

The story is more or less the same on both continents, separated
by a few millennia, but we have come to diametrically opposed
opinions. Faced with one of the biological inheritors of the human-
dominated world, we are left in a quandary that leads to different

reactions. Yet this is the world we now inhabit – one that contains winners as well as losers.

Later in the same summer that my father and I evicted sparrows from the eaves of our family home, I found myself inspecting a bird that was sitting next to me on a stone bench in St Mark's Square in Venice. Ignoring my familial duties, which were to admire stone arches and mosaic-filled churches and attend demonstrations of glass blowing, I turned my attention to Venetian sparrows. The sparrow in question was after my crumbs, competing for food with jostling crowds of feral pigeons – another biological success story. Proudly brandishing my first 'child's bird guide', I set to the task of trying to identify it. It did not look quite like any of the illustrations, but my mother and I decided that it must be a rather unusual-looking Spanish sparrow on account of its rich-brown head. I was delighted to be able to add a new species to my list. When I consulted a more complete 'adult' bird book back home in England, I discovered that we were wrong. Under 'house sparrow, *Passer domesticus*', I found the following words: '*Male Italian Sparrow . . . has brighter coloration in breeding plumage, with rich chestnut crown, whiter cheeks and under-parts*'.[9] There it was, exactly what we had seen – an Italian sparrow. Back in 1967, I was cross that I had to remove 'Spanish sparrow' from my list of birds, and even crosser that I could not add 'Italian sparrow' instead. According to the book, it was a type of house sparrow.

What I did not know as I watched that Italian sparrow pecking at the fragments of the Milan biscuits that I had deliberately dropped was that the sparrows themselves held the secret to their origins in their genes. Just as the genetic 'fingerprints' of humans have on occasion been used to identify the true biological parents in cases of disputed inheritance, the ancestors of species whose origins are disputed can also be analysed. In this case, it should be possible to tell whether Italian sparrows are most closely related to Spanish sparrows or to house sparrows. To find out, nearly fifty years later I set off to visit a group of sparrow geneticists. My route took me through Oslo's Frogner Park, past a few more of the world's half-billion house sparrows and quarter-billion town pigeons, all enjoying a day of

October sunshine. They were picking up scraps from human visitors who were there to see the life's work of the sculptor Gustav Vigeland, a man seemingly obsessed by curvaceous nudes. Eventually, I located the University of Oslo and the building that houses the renowned Centre for Ecological and Evolutionary Synthesis. The door was open, so I just walked in, past cabinets filled with the stuffed bodies of zoological specimens. With no reception to be seen, I carried on, wandering down poorly signposted corridors. Eventually, I arrived at the door, knocked and entered. There I met the guitar-playing, curly-haired Glenn-Peter Sætre, my former student Richard Bailey and several of their colleagues.[10] Glenn-Peter had put together a veritable army of sparrow geneticists, who had set about trying to uncover the history of the house sparrow and the genetic origins of the puzzling Italian sparrow.[11]

They discovered unique genetic sequences, some of which could only be found in Spanish sparrows and others only in the house sparrow – clearly these two were separate species. This was interesting, but not that surprising. The house sparrow had developed in Asia, while the Spanish sparrow lived predominantly in Europe. The surprise came when they analysed the Italian sparrow. The Italian sparrow, it turns out, contains a good old mixture of both Spanish and house sparrow genes. The only logical explanation was that they were hybrids. Glenn-Peter and his colleagues suppose that, as the house sparrow followed the development of villages and agriculture and spread out of southern Asia some thousands of years ago, it met up with the Spanish sparrow. Despite its modern name, the Spanish sparrow lives throughout the Mediterranean region and is not confined to Spain – it must have been found in suitable places in parts of the Italian Peninsula before the house sparrow turned up. It transpired that the hybrids between these two types of sparrow were fertile, and the Italian sparrow was born, with its chestnut-brown cap and white eyebrow borrowed from the Spanish sparrow atop the livery of a house sparrow. A new genetic race had come into existence.

This does not in itself tell us whether we should think of the Italian sparrow as a variety of Spanish sparrow (as my mother and I had supposed), as a kind of house sparrow (which our bird book had suggested), or as a separate species in its own right. In any event, more work was required. Glenn-Peter and his band of merry geneticists found that the Italian sparrows hardly ever still interbreed with the Spanish sparrows, and they remain quite distinct from house sparrows, despite the fact that some hanky-panky does continue to take place in the foothills of the Alps. The Italian sparrow is a new species – separate and genetically self-perpetuating.

Biologists usually think of new species taking hundreds of thousands, or millions, of years to evolve. Yet the Italian sparrow must have come into existence within the last eight thousand, after agriculture first began to be practised in the Italian Peninsula[12] and the previously Asian house sparrow established itself in the region. This is astounding. The expansions of town pigeons, eastern bluebirds, house sparrows and tree sparrows, which are the urban sparrows in parts of eastern and southern Asia, are impressive enough – many previously existing species are benefitting from human-caused changes to the world. But the Italian sparrow is something different. It is an entirely new species that has originated in the human era, and for which we are directly responsible. This species exists only because humans created the towns, villages and farmland that allowed the Asian house sparrows to spread, permitting them to meet up with Spanish sparrows and produce a new kind of hybrid. Moreover, humans have created all the urban and agricultural habitats where Italian sparrows now live. Because hybridization between the house and Spanish sparrows must first have happened in a single breeding season, the birth of Italian sparrows would have been extremely fast. Separation is likely to have been achieved in a few hundred generations, and it may only have taken a few decades, which is at least a thousand times faster than would be expected by our 'conventional' understanding of evolution. It is the virtually instantaneous biological genesis of a new species.

Humans have not only changed the world's ecology, enabling house sparrows to spread around the world, we have also altered the trajectory of evolution.

Amid all this change, our attitudes are failing to keep up with the reality of the modern world. The news is full of stories that we are causing the loss of species and that we are transforming the Earth. As I write, in December 2016, the recent news contains dozens of stories explaining how we are harming the world's biological diversity, pointing out that climate change is causing wildlife populations to die out; that 7 per cent of all mammal species, including giraffes, are endangered by poaching; that thirteen recently discovered bird species are already listed as extinct; that Mexico's vaquita porpoise is close to extinction because they drown in fishing nets; that half of the shark and ray species in the Mediterranean are endangered; and that plastics threaten marine life throughout the world.[13] Stanford University professor Paul Ehrlich sums it all up: '*We are basically annihilating the life on our planet.*'[14] Bombarded by thousands of bad-news stories every year, we are motivated to try and fight back. It is right and proper for us to attempt to stop climate change, to prevent unsustainable hunting and fishing and to avoid polluting our environment.

On the other hand, we are often ambivalent or even hostile to the biological successes of the human epoch. When successful species turn up in new locations, we resist their arrival. The same trawl of current news revealed over a hundred reports highlighting the 'negative impacts' caused by mammal, insect, snail, mussel, worm and plant species that are today living in a continent, country or habitat they did not previously occupy. We have declared them to be in the 'wrong' place, but nearly all these examples could be rewritten as biological success stories. Each of these species is now more numerous and widespread than it used to be. We take a particularly dim view when successful 'foreign' species have the temerity to interbreed with 'native' residents. For example, the European Commission, the UK government and the Royal Society for the Protection of Birds, among others, have been behind recent attempts to exterminate the 'foreign' American ruddy duck, which escaped from British wildfowl collections in the twentieth century and then started to spread into mainland Europe. Ruddy ducks can hybridize with the rarer 'native' Eurasian white-headed duck, so the consensus in the conservation world has been that all the ruddy ducks and hybrids in Europe must be killed.[15] In this quarter, hybrids are deemed to be bad.

Similarly, at the opposite end of the world, pied stilts are elegant, pointy-billed, black-and-white birds that wade through water on long, pink legs in search of tasty insects and worms. Pied stilts colonized New Zealand by flying across the Tasman Sea from Australia in the early nineteenth century. However, they represent a genetic threat to New Zealand's native black stilts, according to the New Zealand Department of Conservation and the International Union for the Conservation of Nature. Hybrid stilts are 'controlled', even though the primary cause for the decline of black stilts has been predation by stoats and other introduced mammals rather than hybridization.[16] Given that introduced carnivores are the greatest threat to black stilts, the long-term effect of hybridization could potentially be beneficial if it acts as a source of genes that enable the hybrids to evade their predators a little better, which the pied stilts evidently can. But the black-and-white hybrids are so obviously 'impure' that they must be removed – it would be a national disaster if the nation's 'all black' stilt were to become extinct.[17]

These human responses to recent biological invasions and to the hybridization that followed suggest that, if the Italian sparrow had arisen within the last few years and had started to spread at the expense of Spanish sparrows, there is a strong likelihood that we would be doing our 'best' to kill off the wicked Asian house sparrow and our guns would be turned on the new hybrid Italian form. It is hard to see the logic of liking Italian sparrows, which are extremely recent hybrids in the history of life on Earth, but being quite so antagonistic towards other hybrids that have even more recent origins. Seen through the telescopic lens of life on our planet, all these events have taken place at more or less the same time. And they all have the same underlying cause: humans.

We have to accept that a world without change is not an available option. Because humans are involved, we feel uncomfortable, giving us a sense of negative responsibility for all the transformations – including new biological successes – we observe. We are confused because a particular animal or plant may not necessarily be where we expect, and confused again because species are neither all good nor all bad from a human perspective. House sparrows are certainly a mixed blessing. They do nest in our roofs, block gutters, eat crops, leave their droppings in stored grain and on windowsills, carry avian diseases and oust other birds from their nesting holes, and they do sometimes peck their competitors to death. But they eat some insect pests and consume weed seeds as well. And, on occasion, we have eaten them as food. They also amuse us. In other words, they are much like any other bird, apart from their particularly close association with humans. They do all these things wherever they live in the world, regardless of how long they have been there. The more recently they arrived, the further they are from their origins and the greater the level of human intervention in getting them to far-flung parts of the world, the more we point to the downside of their existence. But this ignores the reality that house sparrows behave pretty similarly and have comparable effects on humans and other species wherever they live in the world. These different opinions seem to have at least as much to do with human prejudice as they do with any biological reality.

Whatever our individual opinions, these sparrows have been remarkably successful in the modern world. This leads us to an important question. Are they part of a broader movement in which life on Earth has set off on a new course, coping with and adapting to the impacts of humanity? As the remainder of this book will reveal, the answer is 'yes'. Huge numbers of species are thriving in the human-altered world, and we may even be stimulating a mass diversification of new species. The present day is indeed the regrettable end of the evolutionary story for many animals and plants, but it also represents a new opportunity for many others.

PART II

New Pangea

Prelude

The house sparrow reminds us that species are taking advantage of new opportunities that have been provided by humans. Sparrows are ecologically successful in that they are more numerous and inhabit a greater part of the world than they did before humans appeared. They have also evolved – the hybrid Italian sparrow has come into existence, with the consequence that the world's biological diversity has increased by one species. Ecological and evolutionary changes are both of great importance. Ecological success will determine the species that will live among us in the short term, and evolutionary success will alter the future direction of life on Earth. I will consider ecological success in Part II and evolutionary responses to the human-altered world in Part III.

In order to investigate whether opportunists like the house sparrow are rare or common, I have selected four human-caused changes that are critically important to the world's terrestrial biodiversity, and I will explore each in turn. These are the killing of animals for food and other products (Chapter 2), habitat destruction to make space for agriculture and cities (Chapter 3), climate change (Chapter 4), and the biological invasions that take place when species are transported to new parts of the world (Chapter 5). These are the four greatest known threats to the biological diversity of the land but, as we shall see, they have brought unexpected opportunities as well.

There are far too many successful species to chart every one individually, so I have intentionally selected examples from various parts of the world to demonstrate that biological gains are genuinely global. However, we need to complement these illustrative case studies with a broader assessment of how the diversity of life is changing. Is biological diversity going up or down? If we count the number of species on Earth, the answer is undoubtedly down. The 'extinction crisis' is real, as I mentioned in the prologue, and I will return to it later. Nothing that I write in this book contradicts the evidence that we are in the process of losing many species that existed before humans arrived on the scene.

However, it is not so obvious that diversity is declining in any particular district, county, province or nation. Animals and plants are still being extirpated from them, for sure, but fresh immigration provides a counterbalance, introducing new species to each region. We need to enumerate these additions before we can conclude whether biological diversity is increasing or decreasing in the landscapes that surround us. So let us contemplate how we are changing our planet, and tally up the gains as well as the losses.

2

Fall and rise

Ten minutes down a leech-infested forest trail came the first crash. An orangutan, perhaps? More crashes, followed by a hint of musty air. Probably bearded pigs, I thought. Then, rounding the massive buttress of a forest tree were three exceedingly large grey shapes guzzling wild ginger plants and grabbing fruits with their trunks: pygmy elephants. Despite the name, the female was about two metres high and weighed several tonnes, heavy enough to leave plate-sized ponds wherever she trod on boggy ground. Her elder daughter was already a metre and a half at the shoulder. Slightly smaller than their Indian relatives, the pygmy elephants I'd encountered in the Danum Valley Conservation Area of northern Borneo are still among the heaviest animals on Earth. And there were more than three. Surprisingly well hidden by the undergrowth, they were revealed by periodic crashes as they attempted to squeeze through smaller than elephant-sized gaps in the vegetation.

The next crack was much closer. An even larger female was heading straight towards me, oblivious to my presence. At this point, I realized that I had left home without completing a health-and-safety training course on 'what to do if you walk into a herd of elephants in a rainforest'. I moved closer to several large trees, figuring that I might be more nimble than a three-tonne Proboscidean and hoping that the trees would hold their weight (without being at all sure that either was true). At fifteen metres, I reckoned a cough was in order so that she knew I was there. It could have been dangerous if she noticed me only when we were standing next to each other. She registered my cough by raising her head and flexing her ears, but carried on, perhaps imagining that I was one of the herd, or that I was too

31

insignificant to be a cause for concern. At ten metres, I coughed again. Adrenaline coursed through the veins of elephant and human alike. She started, almost jumped, and off she ran: to the extent that a hefty not-quite-leaving-the-ground elephant can run through dense undergrowth.

I realized it was time to head back to the Danum field station for a refreshing drink and to allow my emotions to subside. Safe once more, I was distracted by the Eurasian tree sparrows that had taken up home in this remote outpost, a human-created sparrow oasis surrounded by what for them was inhospitable jungle. They were popping into the canteen through slatted windows and picking scraps of food off plates that had been left stacked up after breakfast. A small bird that had originated in the cool temperate zone of Asia was, thanks to humans, living a mere ten-minute walk from where I had just seen a bunch of forest elephants. This is not how nature used to be.

My only previous encounter with wild elephants was in Pilanesberg National Park in South Africa, where a family party was standing in the middle of the road, barring progress and threatening to leave us trapped in a small car for the night. We would be locked out unless we were able to reach our camp before sundown. The leviathans had seen it all before; tourists were an everyday occurrence for them. Not bothered, they soon lumbered off into the acacia-strewn landscape, and we could speed on towards our lion-free, fortified lodgings. That was more like the wildlife documentary and holiday brochure version of an elephant – African elephants flapping their ears amid savanna landscapes of browned grasslands and scattered trees. David Attenborough might have popped out of the undergrowth at any moment.

The two encounters provided a stark contrast. One group of elephants was travelling in a herd through the basement of a towering, verdant-green tropical forest, the other wandering across sunbaked African grasslands dotted with stunted acacias. Indeed, elephants can be found in near-desert-like conditions as well as in the wet rainforests of the Congo Basin. It raises a question. Why, if elephants can live in such different environments, do they not live all over the world?

In fact, they used to. All manner of elephant-like animals lived across most of our planet's land surface until very recently (in ecological and evolutionary terms), when ancient clans of human primates hunted them to extinction. Elephants could provide food for the entire group, useful skin, fur in the case of mammoths, and a source of bones and tusks to construct small buildings and to make tools. People still kill them today because they are dangerous animals. Exact figures are hard to come by, but the National Geographic Channel documentary *Elephant Rage* reported that they are responsible for the deaths of around five hundred people a year. Testosterone-crazed bull elephants that are in a state of 'musth' during the breeding season are seriously incompatible with everyday life, so Pilanesberg is surrounded by towering electrified fences for a very good reason. But the fences are mainly to protect the elephants, which wander around sporting extremely valuable overgrown teeth. The ivory in these tusks, which is essentially the same as the dentine in our own teeth, has been used for jewellery and ornaments for at least thirty thousand years. Poachers kill the elephants and hack out the tusks, and then criminal gangs ship the ivory to wherever there is a demand, mostly to eastern Asia.[1] It is a deadly battle between humans and elephants, but the odds are stacked: at least fifty times as many elephants have been killed by humans in recent years[2] as the other way around. Worse still, humans can kill elephants faster than elephants can breed. No wonder they have disappeared from most of the world.

The remarkable thing is that any elephants have survived at all – although there is no scientific consensus on why it is the African and Indian elephants that still exist, when the others have disappeared. African elephants had an advantage because they lived alongside pre-human apes from the start and may have gradually evolved an increased ability to escape or fight, allowing them to cope with hunting parties that were armed with little more than sticks and stones. This may also help explain the survival of Indian or Asian elephants, which lived alongside hominids for a million years or more, prior to the arrival of modern humans; although it remains unexplained why it was Indian elephants that survived, when their relatives elsewhere in Europe and Asia did not.

As modern human civilizations developed over the last few

thousand years, the continued survival of elephants seems to be thanks to two further historical accidents. In Africa, they were partially protected by a virtual fence that had been erected by the tsetse fly, a horsefly-like insect that feeds on blood. The painful, saliva-filled bites of tsetse flies transmit trypanosomes, delivering sleeping sickness to human victims, and nagana and surra to our livestock – diseases that incapacitate cattle, camels and horses and afflict pigs, sheep and goats. An estimated 48,000 people south of the Sahara die of sleeping sickness each year, and the toll on our livestock is far greater. This dual assault checked the progress of pastoralists and reduced the encroachment of mixed-agriculture farms into areas where elephants and other large mammals survived.[3] Elephants, in contrast, had a sufficiently tough hide to deter most tsetse flies and enough resistance to trypanosomes to survive in the tsetse zone.

Meanwhile, the Indian elephant seemingly only hung on in regions where they were useful as beasts of burden – in India and Sri Lanka, and eastwards into Thailand and elsewhere in Southeast Asia. They were the living tanks of wars that were fought more than two millennia ago, heavy-lifting machinery for foresters and agriculturalists, and warm-blooded trucks and bulldozers for the construction industry. Wild – or, at least, semi-wild – populations continued to exist in part because it was more practical to capture and then train young, wild elephants than to corral bull elephants and breed them in captivity. Today they are surviving better than their African counterparts because they continue to be used for hauling logs in forestry, and they take centre stage in cultural festivals and religious ceremonies, as well as being used as viewing platforms by tourists. But there is an additional reason. Female Indian elephants and some of the males don't have tusks. They are far less profitable for ivory poachers than their African relations.

Elsewhere, we killed them all, and there are plenty of dead elephants to tell the tale. I even sat next to one, at the Smithsonian Institution's Botanical Symposium recently. This particular African bush elephant was presiding over a collection of scientists and conservationists enjoying their conference dinner in the National Museum of Natural History in Washington DC. A stuffed vulture soared above, forever fixed in flight, attached by a pole high in the museum's rotunda. The unmoving, unblinking elephant looked on; the nearby halls were filled with his dead relatives. The massive skeleton of an American mastodon reminded us that a stroll down DC's National Mall, the green parkland that lies beyond the museum walls, would have been a very different experience in the pre-human past. This particular member of the elephant family had sturdy, curving tusks and a thick coat to survive the cold of North American winters, and they once roamed from where the United States Capitol buildings are situated today to where the Lincoln Memorial now stands, and wallowed in nearby Potomac marsh. This primeval landscape has been replaced by mown grass, planted trees, house sparrows from Asia and North American house finches. Humans from Africa abound where mastodons used to plod.

There must have been twenty or more different kinds of elephants in the world at the time when modern humans began to slaughter them. These included the elephants we recognize today, as well as mammoths, mastodons, gomphotheres and stegodonts, all sharing the characteristics of long, grasping noses and impressive tusks. Only three species survive.[4] The bush or savanna elephant would once have inhabited most of the African continent, from the shores of the Mediterranean to the Cape of Good Hope, but they are now virtually confined to parks, game reserves and, paradoxically, hunting reserves, where they are defended from poachers by wire fences and armed guards. The far rarer African forest elephant is in even worse shape, inhabiting the steamy jungles of the Congo Basin. And the Asian elephants that used to range from close to the Mediterranean Sea to China's Pacific coast are now confined to a smattering of enclaves from India to Borneo. All three are much diminished, but they are the lucky ones.

Seven-tonne, straight-tusked elephants once lived throughout Europe. Elephants and hippopotami would have inhabited swamps at the northern fringes of Europe's Adriatic Sea, now replaced by Venice and its collection of Anthropocene success stories: Italian sparrows, feral rock doves and the inevitable rats and mice. Some elephants had become stranded on Mediterranean islands during the ice ages, when water levels were lower, or had swum across, and there they evolved into miniaturized versions of their ancestors. Malta and Sicily had one dwarf species. Cyprus had another that was just a metre high – a real pygmy elephant. If only they had survived, Mediterranean beaches would have been much more exciting for sunbathing tourists.

We can add mammoths to the list, relatives of the Indian elephant. These include the cartoon-classic woolly mammoth of northern Asia and North America, with its shaggy rufous-red fur coat and arching tusks, and the Columbian mammoth, which lived further south in North America and in parts of Central America. Like the European elephants, island-living populations of mammoths were prone to evolve into tiny versions of their continental relatives. One dwarf island species inhabited the Channel Islands off California, with others marooned in the Mediterranean islands of Crete and Sardinia. Meanwhile, fossils of the American mastodon have been found from Honduras to Alaska, and from Florida to California. The rest of the New World, from Mexico to Argentina, was home to at least three, and possibly five, species of ancient gomphotheres. These animals looked slightly less like modern elephants and were characterized by a protruding lower jaw and extended teeth as well as regular tusks and a trunk.

Perhaps the largest of all was the ten-tonne Asian *Stegodon*, standing four metres at the shoulder, an animal that roamed from Pakistan to Japan, and from China to Indonesia. It is unclear quite how many kinds of *Stegodon* existed, or which of them were wiped out by humans (in several waves of human colonization out of Africa). At least one widespread mainland species disappeared, and a dwarf *Stegodon* lived in the company of dwarf humans on the island of Flores, the elephant surviving until about twelve thousand years ago. Other extinct forms have been reported from Japan and the Philippines, as well as from Timor, Sulawesi, and elsewhere among the islands of Indonesia.

This all adds up to at least one elephant species being present over almost the entire land surface of the world, apart from the driest deserts and highest mountains (if we exclude Australia, Antarctica and the oceanic islands, which they were never able to reach). And they were all killed off by humans.[5] The 'natural' pre-human state of the world is to have elephants virtually everywhere, and it still would be but for the rise of predatory apes.

*

Just a light touch, and the trigger was released. *Thwack*. A beam of the densest imaginable wood crashed down, sufficient to immobilize the tapir between two slanting rows of posts that had been lashed together to prevent the animal from escaping to either side. Rows of posts aligned along the foraging trail of a quarter-tonne Brazilian tapir guide the animal towards its demise. If not crushed immediately, it lies helpless until the spear-wielding Guarani hunters return to complete the task. Fortunately, Gabriel, our indigenous Mbyá Guarani guide to the Atlantic forest at Iguazu in Argentina, was not wielding a spear and there was no tapir. He was simply demonstrating how the largest surviving South American mammal could be captured and killed by a single person. He then continued – with relish – to demonstrate an array of other snares and traps, some baited with fruits to attract gamebirds into a crushing wooden guillotine, others cage-traps that could be tailored to drop on their prey and, finally, the device from Hell – attracted to dangling prey, jaguars trigger a release that deposits the luckless cat on to a bed of spikes. My sister Philippa gasped, as she is prone to do. She and my wife, Helen, had come to admire the splendour of Iguazu Falls and instead were receiving instruction in fatal technologies.

But it was a revelation. It was easy to see how, with several people to operate them, scaled-up versions of the same designs could have been used to trap much larger animals, even those the size of an elephant. And the jaguar trap could have impaled now-extinct, sabre-toothed *Smilodon* cats with little or no alteration – *Smilodon* was about four times heavier than the jaguar and would have declined in number when its prey disappeared, and as a consequence of direct hunting. These traps would not necessarily have killed the largest animals immediately, but they would have held them long enough for armed hunters to move in and complete the task.

Similar Stone Age technologies would potentially have been available to Gabriel's ancestors more than ten thousand years ago, when they spread out across the Americas and encountered large animals that had never met a human before. The wooden traps and heavy beams could have been hewn with stone axes. The ropes, tripwires and bow-strings were fashioned from lianas, sinews and woven fibres. Spears were made of fire-hardened wood to finish the struggling animals off, with arrowheads of glassy stones. It was all simple technology, but lethally effective. None of this required Hollywood visions of muscle-bulging Stone Age hunters locked in mortal combat with fierce beasts. Our planet's greatest animals were brought down more by ingenuity than by brawn.

Had it not been for Stone Age hunters and trappers, we would today be taking our wildlife safaris in Argentina, where gomphothere elephants were joined by huge ground sloths, as well as car-sized armadillos and hornless, rhino-like toxodons. At Luján near Buenos Aires, thirteen different kinds of mammals averaged over a tonne in weight at a single locality.[6] None of them survives. In contrast, the only tonne-sized mammals found on the entire African continent are the two elephants, hippos, white rhinoceros, black rhinoceros and four giraffe species.[7] Game viewing in South America would have been far more spectacular, but not unique. Wherever in the world you are reading this book, you would once have been surrounded by an impressive array of staggeringly large animals.

The forests of New Zealand have lost their three-metre-high giant birds, and the outback of Australia no longer has two-tonne marsupials. Madagascar's vast 'elephant birds' and heaviest lemurs are gone. For a trip down memory lane in North America, visit the La Brea tar pits, today surrounded by urban Los Angeles, just a few blocks from Hollywood's Sunset Boulevard. It is a macabre pre-human animal trap, but a godsend to biologists. Massive ancient animals became stuck in adhesive geological seeps of crude oil and bitumen, and then the predators and scavengers that attended their festering bodies suffered the same fate. The array of unfortunate animals and their predators that have been preserved in the goo reads like fiction: American mastodon, Columbian mammoth, Mexican horse, California tapir, three varieties of ground sloth, American camels, large-headed llamas, over-sized peccaries, enormous short-faced bears, dire wolves, scimitar- and sabre-toothed cats, cheetah-like running cats and American lions. Yet they were real. Alas, the ingenious humans that colonized California over ten thousand years ago were more than a match for them.

These extinctions are almost instantaneous on a geological time-scale but slow enough that they are virtually invisible to their

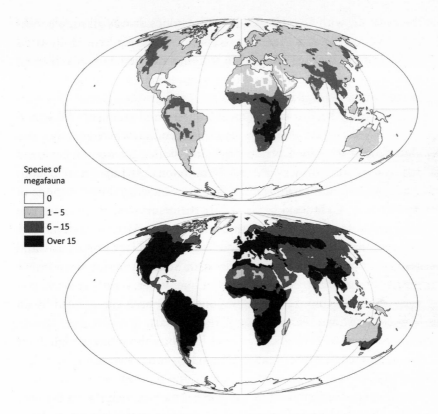

The diversity of wild-living large land mammals, the megafauna (megafauna are traditionally defined as species that weigh at least 100 pounds, or 45 kilograms). Current figures (top) show that the largest numbers of heavyweight species survive in Africa. A reconstruction shows that the whole world would have African levels of large beasts if humans had not killed most of them off (bottom).[8] A few species of extinct bird were also this heavy (particularly on Madagascar and in New Zealand), but are not included on the maps. Domestic animals are not shown.

perpetrators. Despite its brains, *Homo sapiens* forgets far more than it remembers. Most cultural knowledge fails to be transmitted for more than a few generations. Small fragments of bone discarded by our ancestors are all that remain to tell us that a deadly combination of human hunters and their kiore rat passengers drove perhaps a thousand bird species extinct from the Pacific Islands in a little over three millennia.[9] Haast's eagle, possibly the largest ever, passed into

the New Zealand Maori legends as Te Hokioi and Pouakai, a predatory bird so enormous that it was allegedly capable of escaping with
a child in its talons – perhaps it did. But the vast majority slipped
from our societal memories. One species of bird was killed off every
two or three years in the archipelagos of the Pacific Ocean, but they
dropped off the radar, failing to be remembered by today's society.

Likewise, present-day Europeans are generally oblivious to the
absence of rhinoceros and lions in their vineyards, most North American hikers fail to wonder why four-tonne mastodon elephants
are missing from the shores of the Great Lakes, and visitors to Yosem-
ite Valley in California are blissfully unaware that three-hundred-
kilogram sabre-toothed cats should be stalking their campsites. All
these animals would be alive today but for our ancestors. The drip-
drip of extinction rarely registers in an individual's direct experience –
who will remember even one hundred years from now that the last
individual of Rabbs' fringe-limbed tree frog passed away on 26 Sep-
tember 2016? We see the state of the Earth as it is, not as it would
have been, had humans not existed. Only now is scientific research
revealing the magnitude of the changes that took place early in human
history.

Our official inventory of human-caused extinction starts in the year
1500. The International Union for the Conservation of Nature (IUCN)
has logged the disappearance of around 2 per cent of all mammal species since 1500, as well as 1.6 per cent of bird species, and 2 per cent of
amphibians over the same period.[10] According to bone and fossil man
Tony Barnosky, from the University of California, Berkeley, a further
178 of the world's largest mammals disappeared long before 1500, and
the number could be higher. If we combine these two lists, 5 to 6 per
cent of all mammal species have already become extinct in the human
era. And these are just the ones we know about. If we add the thousand
missing Pacific birds to the IUCN figures, then at least 9 per cent of all
the different kinds of birds have already gone, and probably more.
These wild species represent a substantial loss of biological diversity,
almost entirely caused by humans. This is not yet sufficient to equal
one of the so-called 'Big Five' mass-extinction events that took place in
the geological past, when three-quarters or more of all species became
extinct. Taking mammals and birds together, we are 'only' about a

tenth of the way there. But the rate of extinction is still exceptionally high. We have triggered a 'mini mass extinction' and could potentially be on course for a sixth big one if humans continue to dominate the Earth for millennia to come. This would be virtually instantaneous on the schedule of geological time.

This loss is devastating but, luckily, it isn't the whole story. After every fall during the history of life there has been a subsequent rise in diversity. The survivors of past mass extinctions in our planet's history formed an unprecedented set of organisms that inherited and diversified in a world different from any that existed before. Consider the last such event, which killed off the dinosaurs when a ten-kilometre-wide asteroid smashed into what is now the Yucatán coast of the Gulf of Mexico.[11] The exact sequence and timing of events continue to be disputed, but it is likely that dust clouds generated by the explosion resulted in severe climatic cooling, the widespread death of plants and the proliferation of acid rain for several years, followed by extreme global warming associated with a massive injection of carbon dioxide into the atmosphere, which heated the Earth once the dust settled. Whatever the precise details, the consequence was that three-quarters of the world's species disappeared, including the dinosaurs.

This explains why the world now contains over ten thousand different kinds of birds and no reptiles weighing more than a tonne, other than a few crocodiles and their relatives. Perhaps because of their feathery insulation and ability to fly away and search out new food supplies, birds were rather better at surviving the huge swings in the Earth's climate that took place 66 million years ago. We usually think of the rise of the birds and mammals as a tale of diversification after they were released from the burden of having to live with gigantic dinosaurs. And there is an element of truth in this. But for this single, chance event, the La Brea tar pits might be filled with gigantic reptile bones rather than enormous mammals. The absence of dinosaurs was an opportunity for others.

However, a substantial increase in our knowledge of fossils and a revolution in molecular biology is changing this perspective. Birds and mammals did not suddenly appear after the asteroid hit – large numbers of them already existed for millions of years before the dinosaurs took their last breaths.[12] They were not simply reclusive creatures scurrying around in the undergrowth hiding from dinosaurs. A time machine trip back 67 million years, a million years before the asteroid hit, would reveal a rather modern-looking set of birds living alongside dinosaurs in forests that were constructed of a wide diversity of quite familiar flowering plants. Just as now, the forests of South America must have resounded with the haunting piping and whistling calls of birds that would have been much like today's tinamous. These stout-bodied birds generally prefer to walk than fly – tinamous are relatives of rheas, ostriches and kiwis – and communicate through dense forest using a cacophony of sounds tuned to penetrate the vegetation.

The birds that lived in the air, forests, grasslands and swamps of the late Cretaceous period, 67 million years ago, were recognizably modern birds. For example, birds that would later evolve into pheasants could already be distinguished from proto-ducks and geese. Of course, some groups have shown spectacular levels of diversification since then, particularly the passerines, or perching birds, which picked up the evolutionary pace 30 to 50 million years after the dinosaurs disappeared;[13] today's flycatchers, songbirds, crows and finches are passerines, and they number about half of all bird species that are alive today. Eastern bluebirds, house sparrows, tree sparrows, Spanish sparrows and Italian sparrows belong to this group, too, and they are all quite modern inventions. Still, a considerable diversity of birds was living well before the dinosaur-killing holocaust took place.

The story of the mammals is similar – nearly two-thirds of the history of mammals took place in the presence of dinosaurs. Having originated about 166 million years ago, mammals lived perfectly happily (except when being eaten) alongside large reptiles for 100 million years. The ancestors of today's egg-laying mammals (platypuses and spiny echidnas), marsupials (including possums and kangaroos) and placentals (most living mammals) had long been separate groups. Some geneticists argue that many different types of placental mammals were already around before the disappearance of the dinosaurs, but a lack of fossils from this time leaves the question unresolved.[14] Similar stories hold for other groups of animals.

Butterflies in the swallowtail family existed 67 million years ago, and they were already recognizably different from other butterflies. In fact, almost all current families of insects existed long before the end of the Cretaceous period.[15] Plant families were even more resilient. Few, if any, major groups of plants died out at this time, even though many individual species disappeared.[16]

That does not mean that animals and plants were unchanged by the new circumstances. Far from it. Mammals and birds certainly filled a void. The maximum body size of mammals gradually increased, and one-tonne animals were again roaming the Earth about 10 million years after the dinosaurs were gone. They continued to grow larger until about 35 million years ago, since when the largest mammals have remained about the size of the weightiest elephants, although relatives of the rhinoceros have on occasion been a little heavier.[17] The evolution of dolphins and whales came later, the oceans having been emptied of their reptilian equivalents: the plesiosaurs, mosasaurs and ichthyosaurs. The Earth was gradually reverting back to its natural state of huge plant-eaters and uncomfortably large predators, a history that goes back for 200 million years. It stayed that way until humans came along and cleared the Earth of its largest mammals in an evolutionary blink of the eye.

The asteroid that hit the Earth 66 million years ago was a calamity. In a matter of years or decades it wiped out three-quarters of the species that previously existed, leaving the Earth to be inherited by animals and plants that were regular creatures, which had already been around for many millions of years before that time. The future was already part of the past, in the same way that house sparrows and tree sparrows were regular animals 'minding their own business' before they found that the new human-transformed world was to their liking. The inheritors of the future Earth are already among us today, just as they were all those millions of years ago. Although some will be temporary good-news stories, perhaps profiting from methods of farming that will disappear within centuries of their invention, those that succeed in the long run can only come through the ranks of the initial survivors. Only they will be able to leave descendants millions of years from now. Life is the story of the winners.

*

Back at the Smithsonian Museum, we consumed the success stories of the human epoch, surrounded by the fossils of extinct dinosaurs, the bones of mastodons and the stuffed bodies of species that still survive. The winners of modern times are in our glasses and on our plates. Living brewer's yeasts are prospering, converting sugars from the fruits of cultivated European vines into intoxicants for our pleasure. Grape vines, whose fruits evolved to be attractive so that animals would move their seeds to new places, have been extremely successful because they attracted humans. First known from archaeological sites eight thousand years ago on Georgia's Black Sea shores,[18] grape vines are now grown from California to China, in Chile and South Africa, and in Australia and New Zealand. We broke bread made from the ground-up seeds of a human-made plant – wheat – which had arisen from the hybridization between different species of grasses from the Middle East. We ate the delicious muscles of sheep, animals that have come to look less and less like their mouflon ancestors, ten thousand or more years since they were first domesticated in western Asia. They now live throughout the world.

For these chosen animals and plants and fungi, we have entered an era of mutual benefit. This might seem odd because we grind the grains of wheat and maize into flour and kill our cattle and sheep to eat their meat. This is hardly to their individual advantage. Yet this is little different from the way oak and pine trees benefit from the presence of squirrels and birds. Bushy-tailed squirrels scamper off with acorns that fall from tall oak trees, hoarding them for future use. The pied tails of Clark's nutcrackers, a kind of seed-eating crow, can be seen disappearing into the distance as the birds traverse deep ravines, each taking a gulletful of whitebark pine seeds to its caches on high ridges in the Rocky Mountains. Later, when only the windswept ridges are free of winter snow or when they need an extra supply of food to feed their chicks, they will return to reclaim their buried seeds. Acorns and pine nuts will die. But the squirrels and nutcrackers will not necessarily remember the exact location of every buried store, and some of these animals will die before they have consumed their entire cache. The oaks and pines sacrifice most of their offspring to the rodents and birds that will harvest them. But, in the long run, the trees benefit because the squirrels and nutcrackers also plant their seeds in places where new trees will eventually sprout. The seeds that survive become the next generation of trees, akin to our own relationship with the crops we grow and the livestock we keep. Most individual crop plants and domestic animals will die to feed us, but humans ensure that some live, and these in turn spawn subsequent generations. The ecological success of humans is down to this arrangement.

The plants may give up their fruits and seeds, but this is all part of the evolutionary deal. According to the United Nations Food and Agriculture Organization (FAO), maize, rice and wheat cover more than a third of all of the world's cultivated land.[19] They are the most successful plant species that exist today. It is entirely valid to think of them as having taken advantage of a gullible primate to prepare the land for them, sow them, fertilize them, ensure that they are free of pests and diseases, and then keep their seed safe until the next generation can be planted. Who is manipulating whom?

It is bonanza time for animals that can consume these human-tended plants. Where wild beasts once roamed, livestock now graze. Around 30 per cent of the world's productive land is covered by pasture, supporting enormous numbers of large mammals. Others just stand around in barns and wait for us to bring them their food. These animals might not be quite the largest that have ever lived, but cows and sheep are considerably more convenient than fields of mammoths. If we think of the entire planet as one big production system, with plants trapping energy from the sun to convert atmospheric carbon dioxide into sugars and thence into more plant material, animals eating the plants, and meat-eating animals consuming the plant-eaters, then the transformation of the Earth is amazing. Prior to this transformation, the million or so humans that were alive would have represented a small fraction of 1 per cent of the combined body mass (biomass) of all of the mammals that roamed the land, but by the year 2000 humans weighed in at about 30 per cent of the biomass of all land mammals, and most of the rest is our domestic livestock.[20] Humans are hijacking over 97 per cent of mammal biomass to our own ends.

Even more startling, the total amount of mammal flesh has increased – meaning that the weight of all individuals of *all* large land mammals added together is over seven times greater than it was

Four of the world's most successful mammal species in the modern era: a human (my daughter Lucy winning the 'child handler of the year' rosette at the village dog show in 2004), our Irish wolfhound, Rex, a skinny horse and a cow. There are over 7 billion humans (primate), half a billion dogs (carnivore), 58 million horses (odd-toed ungulate) and 1.5 billion cows (even-toed ungulate) in the world. Note the fifth successful species, a cattle tyrant (passerine bird) sitting on the horse's back. These birds presumably used to feed by following extinct South American animals around, catching the insects they disturbed. Today they follow domestic cattle and horses, which are even more numerous.

before humans came along.[21] There are about 1.5 billion cattle, 1.2 billion sheep and 1 billion goats alive, as well as a billion pigs, over 130 million water buffalo, perhaps 58 million horses, at least 40 million donkeys, 13 million camels, some 7 million llamas, over a million domesticated reindeer, and so on. As for carnivores, there are over 500 million domestic dogs and a similar number of cats, although this pales into insignificance compared to 7 billion meat-eating humans (an additional approximately 5 per cent of humans are vegetarian).

Not to be outdone, feathered dinosaurs – birds – are also thriving. There are about 22 billion chickens alive and half a billion turkeys. The total biomass of large birds has increased as we provide them with grain to eat, dose them with medicines and protect them from foxes. As a global production system, the present is not a dip in the total numbers or combined weight of large animals. Surprisingly, it is a substantial increase. The natural state of the world – to be full of large herbivorous animals and carnivores that eat them – continues to the present day.[22]

Soon, the total weight of big land mammals will be ten times greater than it was in the pre-human world, as the human population continues to expand and meat becomes an increasing component of the global diet. In an evolutionary sense, there are now more copies of the genes of our domestic cows, pigs and sheep than there used to be, just as the number of copies of the genes of humans has increased because we eat them. And these species are geographically more widespread than they used to be. The world map of the current diversity of large mammals should really add humans and domesticated cows, sheep, goats, camels, water buffalo, horses, donkeys, pigs, llamas and alpacas. In most parts of the world, the 'true' current diversity of the megafauna is around seven to ten species higher than depicted. This mutually beneficial 'arrangement' (in terms of numbers of copies of genes, not individual welfare) between humans and our livestock means that the Anthropocene is just as much an age of mammals and birds as it ever was.

The extinction of wild species and the rise of domestic animals mean that hunting is no longer required for our sustenance.[23] Nor is it feasible. Ten thousand and more years ago, a mere one-thousandth of the current human population was capable of eating most species of large mammal to extinction, admittedly over a long period. If it were not for domestic animals, the 7-plus billion of us alive today might be expected to polish off all remaining large wild mammals in about a month. Where hunting large wild mammals for food persists, it is usually unsustainable. Bushmeat hunting in Africa, for example, is continuing to drive numbers down and the rational management of the oceans is still some way off. There is not enough wild meat to feed us. However, habits can be hard to break, especially when traditions become luxuries or are linked to particular beliefs. The Japanese do not need to eat whale meat to survive, and large wild animals continue to be hunted towards extinction for their ivory, horn, as pets (for example, killing the parents to obtain infant chimpanzees), traditional medicines, spiritual pick-me-ups and due to cultural tradition, rather than because we need their essential nutrients.

Then there is the tradition of hunting itself, now a hobby more than a necessity. It is an activity enjoyed by those who appreciate the camaraderie among hunters and dream of a more natural past but who, in reality, are likely to spend their time dispatching commercially raised pheasants and quail. Sport hunting, originally the preserve of royalty, has become a tale of the unspeakable in pursuit of the plump, tasty and reasonably easily shot. Captive-reared and semi-tame Asian pheasants are released into the English countryside so that today's spaniel- and labrador-owning townsfolk can have a 'good' day out with their mates in the surrounding fields.

Yet traditions can change surprisingly quickly, all the more so once people are healthy, wealthy and secure. My great-grandmothers presumably looked fetching wearing the plumes of threatened birds in the Victorian era, and my grandmothers kept warm in fur coats and stoles. By the middle of the twentieth century, my mother wore dyed chicken and pheasant feathers in her hat, rather than egret, but still wrapped up in my grandmother's fur coat on special occasions. My sisters wore rabbit-fur and sheepskin coats in the late 1960s and '70s, but no one in the family would wear furs today,[24] let alone the feathers of endangered birds. This complete change in societal norms has taken place in less than a century. Now, we live in an era where in many parts of the world the killing and wearing of large wild animals is no longer regarded as acceptable, although this social transition still has a way to go.[25]

Take Walter Palmer, the dentist from Minnesota who in 2015 suddenly became one of the most hated figures on the internet because he had shot 'Cecil the lion' with his bow and arrow. As it turned out, he was taking part in a legal game hunt, and hunting preserves do protect a wide range of wild animals and plants that cannot live in regular farmland. Nonetheless, most people who had access to computers across the world regarded it as intolerable; and Walter's sins were magnified because Cecil had been given a name. Online commentators were merciless with their barbed comments, just as he was when he dispatched the cat.

As our nutritional needs are now provided by the crop plants we grow and the domestic animals we tend, our urge to hunt can be replaced by other enthusiasms, including conservation, tourism and photography. Shorn of the need to kill large wild animals for meat, bone, sinew, skin and fur, our slaughter has been reduced to the lowest levels in centuries in some parts of the world, especially in Europe and North America. We increasingly see these animals as a source of enjoyment and recreation, and conservationists have been able to enshrine their protection in law.

In the United States, the birth rates of wolves and grizzly bears for the first time in many generations exceed the rates at which humans kill them – both of these rather dangerous animals again stalk the land, recolonizing from their Canadian strongholds. American bison is off the threatened list. The European bison is up from one wild population to thirty-three; the goat-like southern chamois has increased five-fold since 1970; there has been a 14,000 per cent increase in European beavers since the 1950s; and deer and wild-boar numbers have quadrupled in Europe.[26] They are chewing and nose-ploughing their way through the vegetation to the point that conservationists have begun to discuss the 'deer problem' and the 'wild-boar situation'. Their predators are needed to control them.

No problem: bears, wolves, golden jackal, wolverine and lynx are repopulating Europe too – save for Britain, separated by a barrier of water from continental Europe. Wild geese, swans, storks, herons and

A love of fur: guests at the wedding of my aunt and uncle Ruth and Harry Downing, in 1951. Relatives of the groom attended in wild-cat pelts, probably ocelot (top), while those of the bride wore the odd feather with their mink and muskrat coats and stoles (below). In times of post-war austerity, my mother, Diana (bottom right), borrowed her mother's over-sized coat for the occasion, and my eldest sister, Philippa, and brother, Jeremy, had been bought coats with arms that they would be able 'to grow into'. Their descendants in the twenty-first century would not contemplate wearing furs. This change in attitude has enabled the numbers of large mammals and large birds to start to increase again in Great Britain, and in other countries where they are no longer persecuted.

cranes are returning as well, and the great whales, the largest animals ever to have lived on Earth, are once more plying their way across our seaways in growing numbers after centuries of unsustainable butchery. Once we stop killing them, large animals come back, rejoining the 90-plus per cent of smaller ones that never disappeared in the first place. Meanwhile, our domestic animals have reverted to the wild. Horses roam across North and South America. Water buffalo and camels populate the outback in Australia. Wild pigs are common in parts of the USA, and feral cattle populations have established themselves in the Hawaiian Islands, the Aleutians, Campbell Island, and elsewhere. The circle is nearing completion. First, we killed off most of the largest animals. Then we domesticated some of the survivors. And now that we rely on these domesticated animals for our food, it is possible for the still-surviving large animals to prosper once more. Unless some calamity befalls the world (which is entirely possible), there will be considerably more large wild mammals in existence one hundred years from now than there are today.

Where large animals have not returned fast enough for our liking, we have begun programmes of affirmative action, releasing wolves in the wilds of Yellowstone National Park in the States and beavers in Britain's manicured river valleys and ponds. South Africans created Pilanesberg National Park as a deliberate, post-wildlife-apocalypse haven for large animals, a game reserve to service the tourist mecca of Sun City. Humans and their livestock were removed, their villages and farmsteads erased, waterholes added and the whole landscape secured behind a shocking fence. 'Operation Genesis' involved the release of over six thousand individual large mammals into the park in the 1980s: an early example of what is now known as 'rewilding', which is the idea that wildlife can be put back and then left alone to thrive. However, it is not quite so simple. The fence needs to be maintained, the elephants and rhinos guarded, and it was later felt necessary to remove the (previously introduced) wild dogs to allow their prey to survive. Large animals may be thriving once more, but often within the confines of pseudo-wild worlds.

As long as the human condition can be improved throughout the world (sufficiently for us to care more about saving than killing wildlife), such recoveries are likely to become more widespread – at least within human-managed enclaves. Yet, somehow, we yearn for something a little more natural: places that have always been truly wild, places where gargantuan animals still rule and where radical intervention is not required. Places like Danum Valley in Borneo, where I got caught in the middle of an elephant herd. But this sense of a wild world without humans is a mirage. We have transformed the whole planet.

The rhinoceros that used to live in Danum have been hunted to extinction – the last one removed for its own protection – and the overgrown pygmy elephants that live there in their stead are not even native to Borneo. In 1750, the British East India Company gifted some of these great beasts to the Sultan of Sulu, who used to rule the north-eastern corner of the island of Borneo. The pygmy elephants that now live in Danum's jungle are presumed to be descendants of these domestic animals because there are no well-substantiated remains demonstrating the historical presence of wild elephants on the island in the last ten thousand or more years; and 'wild' pygmy elephants are virtually confined to the parts of Borneo where the sultan once ruled.[27] The native languages even lack a word for them.[28]

So, here we have an Anthropocene conundrum. Borneo's elephants were almost certainly introduced a couple of hundred years ago, they eat and squash a lot of native vegetation, cause trouble when they emerge from the forest into oil-palm plantations and farmers' fields and leave huge mounds of dung whenever they walk on the roads. Given these possible 'negative impacts', it could be argued that Bornean elephants meet the international definition of an 'invasive alien species', which means that they should, in principle, be controlled or removed.[29] The counter-argument is that there have probably been elephants of one kind or another in Borneo in the distant past, that the 'normal' state of the world's vegetation is to have very large animals blundering around in it and that Borneo's proboscideans are genetically unique.[30] The sultan's gift probably originated from an elephant population that no longer exists, most likely from Java. If so, they represent the sole surviving population of Javan elephants. According to Prithiviraj Fernando from the Centre for Conservation and Research in Sri Lanka, they are vital to elephant conservation and must be protected. Elephants also attract tourist income. They simultaneously have both beneficial and negative impacts.

Elephants were introduced, or reintroduced in the case of Pilanesberg, to the only two places where I have come face to face with so-called 'wild' elephants. It seems that it is time to stop yearning for a pristine, wild world. We are living on a fundamentally human-altered planet, and there is no longer any such thing as human-free nature. That stopped millions of years ago in Africa as hominids evolved, over a million years ago in southern Asia once *Homo erectus* arrived, hundreds of thousands of years ago in Europe and the remainder of Asia, where Neanderthals and Denisovans roamed, ten thousand to fifty thousand years ago on most of the rest of the continents with the spread of our own species, and many hundreds of years to several millennia ago on most of the world's islands. We cannot reverse time. Instead, we should appreciate changes that are positive as much as we regret any losses. Yes, we have caused the extinction of most of the world's largest land mammals but the American bison is back within the confines of where we allow them to roam. Elephants are doing well in Danum. The grey whale is again migrating in impressive numbers along the western seaboard of North America. Deer and foxes are in our backyards. Nature is fighting back.

3

Never had it so good

The hares and rabbits are nowhere to be seen, perhaps on account of the buzzard hunched up on the fence post and the lingering scent of the fox that visited in the night. The rabbits rarely venture far from the security of their underground burrows. The fresh droppings of brown hares adorn the summits of old ant mounds, but they hide in the long grass in the daylight, avoiding the preying eyes of the buzzard. The buzzard is more interested in earthworms, or at least accepts that worms are easier to catch. Wagtails and starlings accompany ponies in the meadow; black and enamel-red burnet moths zoom past on a blur of wings; pink clovers grow in profusion; damselflies navigate the reeds like miniature helicopters; hawthorn flowers cover the hedgerow with their musky scent; scarlet poppies shimmer in the barley field beyond. It is a quintessential rural British scene, although it happens to be on my own land.

There is nothing particularly unusual about any of these species. None of Britain's rare butterflies live here, nor do any of the rarer dragonflies or damselflies, and there are no unusual bumblebees. Instead we have the sorts of species that you can see anywhere in the countryside. Some of the birds are scarce, but not many of them. Half a dozen tree sparrows, quite rare in Britain but abundant human companions in Tokyo and in the towns and villages of Borneo, are chirping in a bush. As for the ponies, wild horses may be endangered these days but 58 million of their domesticated descendants are still alive. Scanning my land, a naturalist might glance and then move on: these are just ordinary species that you might find anywhere else in Britain.

So, how many are there? I have counted twenty-two butterfly

species, fifteen types of dragonflies and damselflies, ninety-two bird species and nineteen different kinds of mammals living on the land, or at least visiting occasionally. For each of these groups, these numbers amount to between 30 and 40 per cent of all the species regularly found in Britain, which is remarkable, considering that the area of the land is only just over two hectares, or 5.5 acres in old money. That is to say, a third of all British species can be observed in a plot one-millionth of the area of Britain. If we want to go global, four out of every thousand species on the planet can be seen on a billionth of the world's land surface. What is odd about this is that all these species are living in a thoroughly human-changed landscape that bears almost no resemblance to the pre-human habitats that used to be here.

Instead of primeval forest inhabited by straight-tusked elephants and rhinoceros, I have hedgerows, a shelter-belt containing a mixture of tree species that I planted in the year 2000, three small meadows that are grazed by ponies and cut at different times of the year, some bushes, an occasionally dry ditch, a regularly mown lawn, flower beds and three small ponds created using butyl liners. The surrounding countryside is the Vale of York landscape of wheat, barley and oilseed rape. It is a human-dominated world. It seems quite extraordinary for such a small piece of land that is so unlike any 'natural' vegetation that might once have existed in the region to support such a high proportion of the country's animal species, and a surprisingly high proportion of the whole world's species. And few of them are the same species that would have lived in the original forest. This corner of Yorkshire is populated by opportunists of the Anthropocene.

Before drawing any general conclusions about the extent to which species are surviving, and thriving, in human-altered habitats, we need to contemplate other parts of the world. And where better to start than in a tropical forest? A blur of red, yellow and blue stands out in sharp contrast to the green forest trees, as scarlet macaws squabbling over the flowers of a balsa tree screech like drunken teenagers. Capuchin monkeys cautiously reach to pluck fruit from the slender branches of nearby guavas, caterpillars of brightly painted *Heliconius* butterflies nibble tendrils of granadilla passion vines whose exquisitely purple-filamented flowers hang like aerial jellyfish, and banana-like clumps of 'lobster-claw' plants glimmer and buzz with hummingbirds. Malodorous peccary herds pass by, sniffing out the fruits that the capuchins dropped, while lolloping cane toads guard the toilet block. I observe this tropical scene from the comfort of a deckchair, sitting on the veranda of the Sirena ranger station in Corcovado National Park in Costa Rica, one of the world's great wild places. It's an ideal spot to sit and admire the wildlife, sticky-tape in hand, dabbing off the ticks acquired from a day in the forest. It is hardly surprising that Sirena's 310-strong bird list is over three times longer than my own from home: a full 3 per cent of the world's bird species in one place. The park's 140 mammal species also constitute some 2.6 per cent of the world total. It seems like an almost unnecessary diversity, in an apparently more natural environment.

At the height of the wet season in November, Corcovado is humid enough for your clothes to moulder and smell as you suspend them under the roof in the vain hope that they might dry, which I did when I became stranded there just over thirty years ago, an unshaven, mirror-deprived graduate student with the wispiest of beards. On this particular visit, bottle-green wild muscovy ducks paddled in the shallows of the waterlogged landing strip, the deluge rendering Sirena's airborne supply route unusable for weeks. The trail to Los Patos was a twenty-kilometre quagmire, impassable for the park's ponies, which stood forlornly, soaked, licking their wounds. The vampire bats had been at them again in the night. The rivers along the coast were in full spate and blocked our way to the north, but we did not fancy a trip south, where bands of illegal gold miners were in occupation. With all exit routes barred, we sat it out, 'we' being a small group of PhD students from the University of Texas, as well as Don Nilo and several other park guards. We could cope for a while, on a diet of rice, beans and spaghetti, until the beans ran out, and then the spaghetti, and then it was white rice alone. As our energy levels dropped, we entered a state of diet-induced lethargy. Finally, one of our number, Taiwanese student Peng Chai, had had enough.

Starting at dawn, Peng's first task was to catch insects to feed to his hand-reared jacamars, the subject of his doctorate. These astonishing birds sport a coat of shimmering, metallic green feathers and a rusty belly. Their outsized tweezer-like bills are used with great dexterity first to capture flying insects and then to disarm wasps of their stings, to remove the indigestible wings of dragonflies and to inspect butterflies before deciding whether they are sufficiently tasty to consume. With his avian babies fed, Peng marshalled the troops, handing out instructions for gathering fruit: guavas, two types of passion fruit, oranges, plantains, palm fruits and relatives of the chilli pepper of dangerous potency. There was also a relative of black pepper to be found growing beneath the forest canopy, ginger to be dug, crayfish to be netted from the stream, snails from rock pools and fish from the Rio Sirena, where you had to dodge the odd crocodile and bull shark. Add to these ingredients Taiwanese culinary skills, and an array of exciting concoctions was spread before us. Our spirits were lifted, like a miracle.

None of us did any research that day, mind you. So, how did a bunch of inexpert foreigners do, when asked to become hunter-gatherers for the day? We all knew the forest well, but our lack of expertise was all too apparent. Peng and his motley crew managed to gather enough food to repay the energy of collecting it, but only because there was a ready supply of guavas, granadilla passion fruit, citrus, plantains, ginger, palm fruits and coriander, all remnants of the former life of the Sirena station. We were living in a forest that used to be a farm.

Sirena brims with wildlife, but it is not pristine. A previous owner of the main finca – the wooden, colonial-style farmhouse that subsequently became the park-ranger station we inhabited – had kindly planted out most of the bounty that we harvested. Sirena's abundant wildlife was living in a landscape with a past history of hunting, active farming, livestock grazing and cutting of timber. Fewer of those fruit trees survive today as the farmland is gradually swallowed and reclaimed by the encircling forest, but the vegetation and its associated animals reflect this history.

Wild passion vines that benefit from extra sunshine at the human-cut forest edges are sought out by butterflies looking for somewhere to lay their eggs. Fruits of the granadilla passion vines, which had initially been fostered by the farm, are consumed by agouti, long-legged rodents with exceedingly sharp teeth. The planted guavas attract monkeys, peccaries and eye-spotted forest butterflies that suck juices from fallen rotting fruits. Fast-growing balsa trees spring up on the farm's former pastures, feeding the macaws, beneath which coral-orange lobster-claw flowers form an iridescent under-storey, whirring with hermit hummingbirds. The crumpled leaves of willowy, light-demanding *Cecropia* trees that sprout by the side of the human-maintained runway are unfurled by squirrel monkeys seeking tasty insect morsels. Some species, like the macaws, just visit to snack in the former farmland, but the squirrel monkeys, some of the wild relatives of capsicums and their associated clear-winged butterflies, and striped zebra butterflies that are beloved by keepers of butterfly houses in North America and Europe, are barely seen at all in the denser forest beyond. Numerous species owe their presence – or at least their abundance – to the habitats created by the previous human presence.

There is a long history of human occupation in this region. Up to two metres wide and fifteen tonnes in weight, the remarkable stone spheres that were left by the ancient Diquís culture were hewn from dense gabbro rocks. No one knows quite why they were created, but whatever their original function, they speak to us of a pre-Columbian culture in this part of Costa Rica that was capable of monumental feats. These feats included farming the river valley directly inland from Corcovado National Park for a thousand or more years.[1] The Diquís bordered on the maize-growing Aguas Buenas and, later, Chiriquí cultures of western Panama and southern Costa Rica, which were established from AD300 onwards. Farming is not a recent event in this part of the world, or in most other places. Far from being a pristine rainforest, the human-associated wildlife that we enjoyed at Sirena has been adjusting to disturbance by humans for thousands of years.

The same is true elsewhere. Agriculture developed in Mexico as early as ten thousand years ago. Over the following several thousand years, cultivation was established throughout Central America and in a band across the northern parts of South America.[2] By six thousand years ago, agriculture was practised in the Andes and southern Amazonia. Millennia later, the conquistadors rode through farmers' fields more often than they hacked their way through virgin jungles. Tilling the land and pastoralism began between thirteen thousand and ten thousand years ago in the Middle East, and has also been widespread in Europe and northern Africa for many thousands of years. Cultivation of broomcorn millet started in the Huang He basin of China around ten thousand years ago, and the inhabitants of Southeast Asia later transformed entire landscapes into terraced wetlands for the cultivation of rice. Further east, the mountain tribes of New Guinea have cultivated their farms and gardens for nine thousand years, surrounded by the forests where they still hunt. In Africa, too, agriculture has a long history. Ethiopia was home to a succession of humanity's distant relatives, culminating in the evolution of our own species two hundred thousand years ago. Despite millions of years of hominid history, and despite having been an important centre for the domestication of crops for thousands of years, Ethiopia is still rich in animals and plants found nowhere else on Earth.

Not only are these transformations ancient but they are truly global. North American hunters used fire to manage forests and grasslands, as did Australian Aboriginals. The coastal forests of Okomu National Park in Nigeria are strewn with historical artefacts of previous human occupation; what was once thought to have been pristine forest has regenerated in the past seven hundred years, growing on a bed of charcoal. Fast-forward to soya farms in Brazil, agricultural prairies in Kansas, wheat fields in parts of Russia, enormous Australian sheep paddocks and silage production in sown pastures in northwestern Europe. Humans have created a cornucopia of new habitats, and have done so for millennia across six continents. This has created a world of new opportunities for those animals and plants capable of seizing them.

That so many species now live alongside us in human-modified environments, whether on intensive farmland in Britain or in regenerating forest in Costa Rica, is not to say that any human society, past or present, has ever lived in 'harmony' with nature. This is absolutely not the case. The harmonious coexistence of humans and the rest of nature in the distant past is a romanticized and largely fictional notion. Present-day conservation often attempts to re-create these idealized ecosystems, for example by the reintroduction of hunter-gatherer-style burning of vegetation in America, Africa and Australia, and by reinstating now uneconomic medieval farming and forestry practices in Europe and Asia. In truth, our relationship with nature is, and always was, less romantic. We eat nature. We take up space that wild nature would otherwise occupy. We have used whatever technologies have been available to us at a particular time to consume or oust wild creatures, often with great success. As a consequence, we are living through a time of extinction.

It is important to consider these losses of wildlife, particularly those associated with the conversion of so much of the world's surface to human ends, if we are to put the successes of the opportunists into perspective. Once we convert the previous vegetation to cereal fields and concrete, there is less space left over for the rest of nature. One way to estimate the impact of these changes is to apply the so-called 'species–area relationship' to the loss of the original habitat. This relationship tells us that the number of species declines when the area of any habitat is reduced; for example, relatively few forest birds live on the smallest wooded islands that can be found off the coast of Brazil.[3] These islands became isolated from the mainland when water levels rose at the end of the last ice age, 11,650 to 7,000 years ago, and so they provide insight into how many forest-dependent birds might survive in the long run, if we were to convert a continuously forested landscape into one in which there are 'forest islands' surrounded by an ocean of sugar-cane fields and pastures. This is, of course, exactly what we have done. The species–area relationship tells us that we can normally expect somewhere between 29 per cent and 44 per cent of species to disappear if 90 per cent of the original habitat is destroyed.[4]

Armed with such species–area graphs, Tom Brooks from the International Union for the Conservation of Nature and Andrew Balmford of the University of Cambridge took on the challenge of estimating how many species might become extinct because of forest loss in the Atlantic region of Brazil. These Atlantic forests form a wet coastal fringe from the north-east to the southern limits of Brazil, extending into northern Argentina and Paraguay, and separated from the humid Amazonian forests by the dry, savanna-like 'cerrado' vegetation, which is inhabited by termite-eating giant anteaters and maned wolves. The cerrado is too dry for most Amazonian and Atlantic trees, so these two great forests and their inhabitants are kept apart, their long history of separation allowing the animals and plants of coastal Brazil to evolve into species that live nowhere else in the world. The golden-lion tamarin, woolly spider monkey, buffy tufted-ear marmoset, yellow-breasted capuchin, Barbara Brown's titi, Southern brown howling monkey and Northern muriqui head up a long list of monkeys that are entirely confined to these forests. It is also the home of most Brazilians – São Paolo and Rio de Janeiro are bang in the middle of it. Not surprisingly, therefore, some 93 per cent of the forest has been cleared, principally for agriculture, and the region holds one of the world's greatest concentrations of endangered species.

Brooks and Balmford focused their research on the forest birds rather than the monkeys. They estimated the amount of forest that had been lost in each part of the region and then used the species–area graphs to work out how many of them might be doomed.[5] Their answer was that 51 (or 41 per cent) of the original 124 forest-dependent birds that they looked at could be expected to disappear. Of these 51, some 45 have already been designated as globally threatened with extinction.[6] Brooks and Balmford concluded that, without remedial conservation actions, all of the threatened species might be expected to die out over the coming decades and centuries.

One already has. The Alagoas curassow was previously found in the north-easternmost part of the Atlantic forest region. This fruit-feeding, glossy-black creature is reminiscent of a slimline turkey but with a prominent and curiously flattened red beak that gives it the appearance of having had its snout trapped in a door. Its misfortune was to have lived in Alagoas, the most densely populated area of Brazil, aside from the metropolises of Rio, São Paulo and Brasilia. Consequently, all but the very steepest 2 per cent of the former forested land in this region has been converted into sugar-cane fields, tropical farmsteads and urban developments. This misfortune was compounded by the fact that this particular bird makes a delicious meal. Luckily, around 130 individuals still survive in two small captive populations, albeit with a hint of razor-billed curassow genes in the mix – an attempt to restore genetic variability to the inbred remnant population. Nonetheless, it is extinct as a truly wild species.

The Alagoas curassow was not the only one, or so it was thought. The other candidate to have joined the dodo on its path to extinction was the small green-and-yellow kinglet calyptura. According to English ornithologist William Swainson's charming nineteenth-century painting, this bird was characterized by a dashing topknot of rusty-red feathers. Unseen for a hundred years, it seemed reasonable to assume that this species was also extinct when Brooks and Balmford penned their article in 1996. The birds themselves were entirely oblivious to deliberations about their fate, however. As if both to annoy (by making their article immediately slightly out of date, though not in any important way) and to delight (because it was still alive) the authors, the calyptura revealed its presence once more in the very same year that Brooks and Balmford published their work. Nonetheless, the curassow and the calyptura (which has not been seen since) are under threat, as are many additional species in the region.[7]

The situation is equally worrisome for the other inhabitants. Surveys of forest mammals in one part of the region estimated that 78 per cent of their populations (but not whole species) have already become extinct.[8] A frog seen just once, ninety-eight years ago,[9] is unlikely to enjoy a calyptura-like resurrection, given that it was only ever spotted in the outskirts of urban São Paulo. And at least five plant species are presumed extinct in the region. The threat that Brooks and Balmford sought to quantify across the Atlantic region of Brazil is certainly real.

On the other hand, we need to consider both sides of the equation. When people have counted changes to the number of species of animals or plants in any given location over time, they usually find that local diversity has stayed about the same – a conclusion that is based on the analysis of large volumes of data that have been collected over many decades.[10] If anything, the average number of species is increasing slightly. That must mean that new species are arriving at least as fast as any previous occupants are disappearing. And crops, pastures and urban environments still average about 60 per cent as many species as can be found in equivalent areas of the habitats that preceded them.[11] The original habitat is not so much destroyed as replaced by a new environment that still contains quite a lot of species. Once one appreciates that there may be several different human-created habitats in any given region, containing somewhat different species (i.e., the species found in crops, pastures and urban areas are not all the same), then the total number of species found within a region may be just as high, or even higher, than it was before.

In the Atlantic region of Brazil, only one of the forest birds has died out and over half of the remainder are not threatened. These survivors are successful species of the Anthropocene epoch by virtue of their capacity to tolerate living in small areas of forest, or to adjust to some level of disturbance. Then there is an influx of new species, which are genuine beneficiaries of humanity. Rusty-white cattle egrets strut in great numbers, hunting for insects and frogs that are disturbed by the hooves of domestic cattle, using skills that their ancestors honed following herds of buffalo and zebra. Originating in Africa, these enterprising herons managed to fly right across the Atlantic Ocean and colonize South America in the 1870s, eventually arriving in São Paulo and Rio in the 1970s. Now, they stalk cattle in Atlantic pastures, joined by yellow-bellied cattle tyrants, ground-dwelling birds that are South American natives but which were formerly excluded by the forest. An ecological auditor might ask: how many species have been lost and how many gained? Alagoas curassow: debit. Cattle egret: credit. Cattle-tyrant: credit. One loss, but two gains – and the list of arrivals goes on. Burrowing owls have invaded farmland that was previously covered by forest, these long-legged hunters liking to chase their prey across open ground rather than to swoop through the trees that used to cover the land. There seem to be more globe-trotting inheritors of this human-transformed part of the world than there are casualties. This suggests that the present-day mixture of small areas of relatively undisturbed habitats and all sorts of human-created environments might in fact contain more species than we started off with.

The antics of natural historians give us a clue that this could be so. In the movie *The Big Year*, Jack Black, Owen Wilson and Steve Martin caricature a real-life story of an obsessive competition to spot more species of bird in one year in North America than anyone else ever has. We birders are a strange lot. We have life lists, country lists, year lists, garden lists and holiday lists (but 'holiday' sounds far too frivolous, so we call them 'trip lists'). When a birder goes on holiday, they aim to visit as many different habitats as possible, including those that have been created by humans. Eschewing another day on the beach, where they face the horrendous prospect of getting sand in the workings of their binoculars, the dedicated birder sets off to sample the delights of the local sewage works. They are off to spot a few waders that are probing the soft ground for invertebrates, or wagtails sallying after juicy flies attracted by the human effluent. The goal is always to maximize the number of different species that can be spotted.

Similar strategies are adopted by plant and butterfly hunters, though only rarely with quite the same level of obsession, and with a

African-origin cattle egrets can be seen coming in to roost at night after they have spent the day hunting for insects on human-created pastures in the Atlantic forest region of Brazil. Despite so many species being endangered by deforestation, the total number of species of bird has increased in this area. More species have arrived to exploit human-created habitats than have been lost to deforestation.

welcome reduction in the emphasis on sewage. The naturalist's desire to visit different types of habitat implies that the total number of species in a mixture of ecosystems is likely to be greater than if an entire area was covered by one type of vegetation. But are these naturalists correct? If so, the current mixtures of habitats that exist all across the world, which have been modified by humans in a variety of different ways, are likely to contain more species in total than existed in the original vegetation.

If commandeering land for human purposes is going to reduce the diversity of species anywhere, then we should head for a tropical forest. This is where there is most biological diversity to lose. It is where the colonization of human-created habitats by additional species is least likely to compensate for the losses.

After six hours of driving through alternating clouds of red dust and tyre-sucking, muddy pools, arriving at the entrance to Korup National Park in western Cameroon is about as spectacular as you can imagine. A sweeping suspension bridge ushers intrepid travellers towards the forest, capably assisted by the loquacious Chief Adolf of Mundemba's machete-wielding guides. One of the most diverse rainforests in Africa, Korup teems with life: over 600 species of trees and shrubs, 500 herbs and climbers, 1,000 kinds of butterfly, 130 or more types of fish, and some 200 reptiles, frogs and salamanders can be seen. It is home to chimpanzees and to the endangered drill, a relative of the baboon, whose black-faced males sport incongruous white pencil moustaches above their pouting pink lower lips. Both these primates are threatened by the dual pressures of habitat loss and the hunt for bushmeat. The park is also a long-standing home to the Korup people, now relocated into a forest support zone, a buffer on the fringes of the park.

The villagers of Basu, Bajo, Abat and Mgbegati in this buffer zone grow annual crops and maintain agroforests, where scattered trees shade groves of coffee, cacao and plantains. Regenerating 'secondary' forests grow on steep hillsides that have previously been logged, along with 'near-primary' forests, that is, those which are just about as undisturbed as you can get in this peopled forest. Has all this disturbance added or subtracted species? German scientists from the University of Göttingen, working with Serge Bobo of the University of Dschang in Cameroon and Moses Sainge, the park's field manager, set about discovering the effects of these human interventions on the diversity of birds.[12] Rising before dawn, they traipsed out to their field sites to count the number of species, the early rise enabling them to see the birds at their most active, when they are easiest to locate and to identify by their calls. Small parties of the rather dull-green Fraser's sunbird were noticed moving through the branches of the near-primary forest, whereas the shimmering-green and scarlet, olive-bellied sunbird and the iridescent, blue-green superb sunbird staked out flowering trees and shrubs in the more open habitats. What might have seemed like the ornithological short straw of being assigned to count birds in patches of annual crops near to the villages turned out to be a boon. The field workers saw and heard around one and a half times as many individual birds per visit to the annual crops as they did in the forest. And, overall, the numbers of species they observed were pretty much the same in all four types of habitat. The more human-modified habitats contained lots of species – not exactly the same species as in the forest, or in the same abundances, but as many different species. Birds that feed on insects and follow the massed ranks of driver ants (hoping to catch insect prey that the ants scare away) are most abundant in the near-primary and secondary forests where the ants themselves are most commonly found, but the reverse is true for those that eat seeds and for flower-visiting sunbirds. Because some species are found only in forests and others are confined to agroforests and annual crops, the human-created mixture of forest and agricultural habitats thus supports more species than could be found in a forest-only landscape.

Meanwhile, the researchers discovered that butterfly diversity was in fact higher in the agroforests and secondary forests than in the primary vegetation,[13] and more species of under-storey plants were found among the annual crops than elsewhere. Across all groups of animals and plants that they considered, the mosaic of habitats around these villages results in an astonishing diversity. This is not an argument to stop protecting near-primary forest, which contains the greatest variety of trees and many rare animals that are confined to West Africa (whereas the agroforestry and annual crops contain higher proportions of more widespread and successful species).[14] Protecting the largest, least disturbed and least poached forests in this region is the only effective way to ensure that all the species are safe.

But if we try to answer the scientific question 'What happens to the total number of species present when uninterrupted forest is converted into a mixed landscape that contains a patchwork of forest remnants and various human-created habitats?', the answer is that the number of species increases. This is because new species move into human-created habitats faster than the previous residents of the region die out.

Does this conclusion hold elsewhere, in other kinds of habitats? At the opposite end of the ecological spectrum from the rainforests of Cameroon is Flanders, the flat, Dutch-speaking part of Belgium, which was once described by butterfly ecologists Dirk Maes and Hans Van Dyck as Europe's biological 'worst-case scenario'.[15] It is a region of intensive agriculture and densely populated towns and suburbs. It is certainly a good candidate for a reduction in habitat diversity and for more extinctions than gains. And, truth be told, Flanders is not top of my list for a wildlife tour of the world. But if we consider the whole of Belgium, the animal and plant lists for the country include species found in remnants of the original woodlands, bogs, rivers and sand dunes, plus those associated with the human-created rough grazing land, heathland, hay and silage fields, managed woodlands, plantations, horticulture, different types of cereal and root crops, orchards, ditches, road verges, suburban gardens and parks, farm ponds, canals, reservoirs, barns, glasshouses, the outsides and insides of a variety of buildings, brownfield sites and even parking lots. Just as in England, house sparrows, yellowhammers and meadow butterflies can be found in Belgium because rather than in spite of humanity. Add up the species in all these habitats, and the region has a greater diversity than we started with, despite the fact that many of the original species have been lost.[16]

Losses of the original species can be minimized if 10 per cent or more of the land surface remains in a relatively undisturbed condition – ideally, 30 per cent should be protected in some landscapes within each biologically distinct region so as to maintain animals and plants that are particularly sensitive to habitat change.[17] This level of protection is not achieved everywhere, but, even so, most landscapes that today contain a mixture of human-created habitats and at least some remnants of undisturbed vegetation hold more different species than they did before humans arrived on the scene. Perhaps biodiversity has never had it so good. But are we sure about this? A provisional answer was provided back in 1995, when Michael Rosenzweig, a cheery scientist from the University of Arizona, deduced that places with a greater diversity of habitats do indeed contain more species.[18] When investigating the species–area relationship of the type that Brooks and Balmford used to estimate the risk of extinction among Brazilian Atlantic forest birds, Rosenzweig discovered that habitat heterogeneity (the mixture of habitats) was an even better predictor of how many species would be found in a given location. Other researchers have come to similar conclusions, including Jonathan Hiley, who leads a quadruple life as a school teacher, Mexican cricket international, ornithologist and PhD student based at the University of York. He discovered that a combination of human-modified and relatively undisturbed habitats in the Sierra Gorda Biosphere Reserve in central Mexico has increased the number of species in the region.[19] Brown-headed cowbirds and white-winged doves, for example, have moved into the farmland and villages, adding to the area's diversity, while other species continue to survive within protected habitats. Comparable conclusions have been reached for butterflies in Borneo, insects in Hungary and plants in Germany. All these individual studies have culminated in an overarching analysis by Anke Stein and Holger Kreft from the University of Göttingen, together with Katharina Gerstner from the Helmholtz Centre for Environmental Research in Germany. They rifled through the scientific literature from across the world and unearthed over a thousand relationships between the number of species seen and various measures of habitat diversity. The answer was conclusive: the number of species almost always increases with the diversity of habitats.[20]

This is all very well, provided that the diversity of habitats is in fact increasing. There is a common perception that humanity is 'simplifying' nature and reducing 'habitat heterogeneity'. Researchers and conservationists often suggest that we are reducing the variety of habitats in a particular region. But this perception is rarely valid unless we omit human-generated habitats from the equation – which these studies usually do. Of course, there are places where ferrous herds of ploughs, motorized seed drills, fertilizer and herbicide booms and combine harvesters sweep the landscape on such a massive scale that it is hard to find any trace of the former vegetation, as in parts of the Midwest of the USA, and possibly even in bits of Flanders. We are definitely capable of reducing the variety of habitats that can be found, and wildlife diversity will have declined in some of these regions. But we have in fact increased the number of habitat types in most Belgium-, Maine-, Panama- or Lesotho-sized regions of the world. Stein and her colleagues' research shows that this will normally enhance the biological diversity of those regions, not erode it. Now we know this, we can actively increase the variety of habitats and thereby the numbers of species in any given region – if we wish to.

This all means that there are myriad successful species growing, crawling and flying across human-modified landscapes. Agriculture and other human-made changes to the world's land surface are not inexorably reducing all forms of diversity. Most of the species that I have at home in England are there only because they live in habitats created by generations of Britons; large parts of Belgium are just the same; numerous species in the Korup region of Cameroon are found only in places where humans have replaced or disturbed the forest; species are thriving in former farmland at Sirena in Corcovado National Park in Costa Rica; and newcomers to the Atlantic forest region of Brazil have exceeded the number of species that have been lost. It is probably the same where you live too. The number of species in most regions is increasing, not decreasing.

4

Steaming ahead

Three groups of primates were perched above the dramatic gorge of the Jemma River, a tributary of the Blue Nile. Agricultural terraces lined the gentler slopes, interspersed by rugged crags where rusty-breasted, moustachioed vultures soared on the up-draughts. Each group had its own purpose. Droves of smartly dressed, brown-skinned *Homo sapiens* made their way along the paved road, heading on their pilgrimages to the nearby Debre Libanos monastery, seeking absolution from the humdrum drudgery of mortal life. The second group, to which my wife Helen and I belonged, was considerably scruffier; sun-hatted, pink-skinned *Homo sapiens* resplendent in dangling binoculars and telephoto lenses. We scurried in a more chaotic manner towards the cliff edge, engaged in a wildlife pilgrimage. One of the group, a septuagenarian greybeard, was particularly dogged, seeing how close he could get in his enthusiasm to capture a fine portrait of his quarry – the third primate. This turned out to be a few metres closer than anyone else thought was wise. The third group of primates was not afraid of pale, pink humans.

The baboon-like geladas sought their own solace in a bachelor troop, sitting comfortably in the dappled shade of imported Australian *Eucalyptus* trees, sniffing the air for the scent of fertile females. They were relaxing, avoiding the heat of the day, while the somewhat less hairy females and their infants were happily munching their way through a field full of grass; their penchant for eating crops explains why the locals quite reasonably throw stones to keep the greedy animals away. Little did the geladas of Debre Libanos know, but greybeard Roger and his companions had just flown thousands of kilometres in their quest. In so doing, they had just contributed more than their fair share of greenhouse gases to the atmosphere. Our contribution to climate change was potentially more dangerous to the geladas than a few stone-wielding locals.

The mountain geladas are covered in a thick pelt of silvery-brown hair, punctuated by a bizarre, hourglass-shaped pink expanse of chest skin. Their dense fur enables them to cope with cold mountain conditions, but they would overheat – and their diet of grass would be of poor quality during the dry season – at the sweltering lower elevations where olive baboons can be found. Confined to altitudes above 1,800 metres, the only places in the world where geladas can be found are in Ethiopia's towering Simien Mountains and in gorges like that of the Jemma River.[1] Temperatures in mountain ranges decrease by between about 0.5°C and 1°C for every hundred-metre increase in elevation,[2] so we can work out how far they might be forced to retreat as the climate warms. This part of the world may heat up by as much as 5°C over the coming century;[3] if this were to happen, geladas would need to move upwards by 500 to 1,000 metres – potentially to elevations above 2,800 metres. Fortunately, these hairy primates can already be found higher than this in the Simian Mountains, so they should survive global warming, even if populations below this disappear, like those in the Jemma Gorge.

Not so the Ethiopian wolf, squeezed between humanity and the clouds. While the gelada population numbers around two hundred thousand, only five hundred orange and pale buff Ethiopian wolves remain, of which barely two hundred are mature, reproducing adults.[4] They are restricted to much higher altitudes than the geladas. About 113 of these adults, over half of the world population of the species, are confined to the Bale Mountains, and this is where they are easy to see. Or so we were told. Alas, our early-morning departure to watch wolves foraging at dawn was thwarted because a section of the precipitous road to the high plateau had been washed away by a storm in the night.

Abiy, our Ethiopian guide, and an assorted collection of British holiday-makers leapt from the vehicle to gather stones, filling the gullied track with whatever came to hand. An eternity later, or so it seemed, the brave troop of hominids had filled in the deepest of the ruts and our driver could attempt the climb. But alas, we were too

Gelada families (above) graze by picking grass blades with their rather stubby fingers, in a pasture perched above the Jemma Gorge in Ethiopia, where they retreat to steep crags to avoid disturbance (below; notice a few geladas on the rocks in the foreground). Groups of males sit out the midday heat in the dappled shade of Eucalyptus trees (right).

late: the wolves had gone back to bed for the day. Refusing to give up, we spent the following six hours with numbed fingers, dizzily rushing back and forth in the thin 3,500 metre-plus mountain air – but the closest we got to a wolf was to spy a distant orange spot on the horizon. Reluctantly, frozen to the core, we set off home across the undulating plateau when, suddenly, over the brow of a ridge, there it was: a full-sized wolf standing next to the road.

The wolves had come back out again for their evening snack of giant mole-rats, an endangered rodent that is completely confined to the Bale Mountains. It was hard not to feel sorry for them. Like

The Ethiopian wolf (opposite) is confined to the mountains of Ethiopia, with more than half of the world population living in the Bale Mountains (above). The wolves inhabit the freezing-cold Afro-alpine zone, where giant lobelia plants pepper an otherwise tundra-like mountain vegetation. There, the muddy-pawed wolves dig (opposite, top) for burrowing giant mole-rats (opposite, centre), whose world distribution is restricted to the Bale Mountains. Both species are endangered by human-caused climate change.

enormous hamsters or lemmings, they excavate underground passages, only coming up to graze on mountain vegetation that is close

enough to their burrows that they can retreat if danger threatens in the form of a wolf or an eagle. The landscape is full of meal-sized rodents, and this allows the wolves to live at higher densities in this mountain range than they could anywhere else.[5] There are about two and a half thousand giant mole-rats per square kilometre of suitable habitat, which explains why over half of the world population of Ethiopian wolves lives in the Bale Mountains, and why the wolves have muddy paws from digging them out. Although mole-rats can be found from 3,000 metres to nearly 4,200 metres altitude, most of them are restricted to a much narrower band, primarily in the short Afro-alpine vegetation at 3,500 metres and above. This means that they, along with the wolves, are truly endangered by climate change.

The wolf itself would probably be perfectly happy living in the warmer lowlands, but this would bring it into conflict with people or, rather, with their domestic dogs. Dog-borne rabies and distemper have extinguished the susceptible wolves from lower altitudes (there are no recent wolf sightings below 3,000 metres) and periodic epidemics threaten even those on the high plateau. Warmer temperatures may bring people and their dogs to higher elevations and confine the wolves yet more.[6] Perhaps the greater risk from climate change, however, is the disappearance of their food. Move down just a couple of hundred metres from the frigid Afro-alpine zone and one arrives in a shrubbier vegetation where mole-rat and wolf densities are barely a tenth of those on the high plateau. If human-caused climate warming causes this somewhat lower-elevation vegetation to start growing 500 metres to 1,000 metres higher than at present, virtually the whole of the area would turn into a rat- and wolf-poor shrubbery. The maths is grim. A nine-tenths reduction in density applied to the 113 mature breeding wolves that currently live in the Bale Mountains would be 11, and there would probably only be 20 to 30 animals in total if non-breeding adults and juveniles are included. This would not be enough for a viable population to survive several centuries of a warmer climate. Unlike the geladas, there is no opportunity for the mole rats and wolves to survive anywhere higher.[7] The world's climate is getting hotter, and they have nowhere to go.

*

Future threats of the kind faced by the Ethiopian wolf seem clear enough, but we should remember that global temperatures have been rising since the 1970s. Climate change is not only about the future. This gives us the opportunity to find out whether these dire predictions are likely to be borne out. Have mountain species begun their retreats to higher elevations?

In 2007, my fearless PhD student I-Ching Chen set out to find the answer. The task she set herself was to climb up and down Mount Kinabalu in Borneo, lugging heavy batteries and awkward moth traps along with her. Catching moths on a tropical mountain might seem like an odd activity, but moths are lovely, delicately camouflaged animals. Some of them resemble dead leaves, and others the moss- and lichen-encrusted trunks of forest trees, patterns that enable them to remain hidden from the prying eyes of insectivorous birds that seek them out during the day. More to the point, moths form part of the most diverse set of animals on our planet, tropical insects, and hence they represent a crucial element of the world's biodiversity – eating the vegetation and in turn being consumed by larger animals. The choice of Mount Kinabalu itself was dictated by the fact that it was possible to repeat a survey of moths which had previously been carried out by three undergraduate students during their university vacation in 1965. After she and her intrepid husband had spent weeks with aching backs and limbs trapping the moths, and another year peering at the specimens to identify them, I-Ching was in a position to answer the question. The moths had indeed moved upwards, retreating from the lowlands. Moreover, the species that live in the ever-damp cloud forest have exhibited steep declines, seemingly caught in a climatic pincer movement between temperatures that have become too hot at low elevations and changing cloud cover higher on the mountain.[8] The risk of extinction is real.

Half a world away, in the spectacular cordilleras of Central and South America, dozens of species of harlequin frog have already disappeared. They are – or were – among the most beautiful amphibians in the world; their brilliant oranges, blues, blacks, greens, purples and pinks (depending on which species of frog it is) warn would-be predators that their skin contains a deadly neurotoxin. Sadly, these frogs have suffered from a perfect storm of human-caused climate warming, El Niño currents in the Pacific (which accelerate warming and alter rainfall patterns), and epidemics of a nasty fungal skin disease which is thought to have originated in Africa.[9] Each successive hottest year brings new fatal epidemics of the invasive fungus, and yet more extinctions.

The prognosis is not good. Climate-change casualties are set to become the next phase of humanity's mass extinction – at least 10 per cent of all species that live on the land are expected to perish, and possibly double this number.[10] In the face of such dire predictions, what can we do? Of course, reducing our greenhouse-gas emissions is the number-one priority, but even the lowest expected levels of future warming are liable to exterminate many of these species. They are trapped. Animals confined to tropical mountains cannot descend to the lowlands and seek out colder mountain ranges thousands of kilometres away. Insects and plants that live only in moist ravines cannot cross deserts to reach new homes. The only realistic conservation option – if we wish to save them – is for us to start transporting them to new locations where they will find the future climate more to their liking.[11] Yet how do we know that species are more likely to survive if they move to new places? We need to know how species have survived periods of rapid climate change in the past.

When, in 1954, the densely bearded and exuberantly eyebrowed English scientist Russell Coope visited Upton Warren, a damp hole that had been created by the extraction of gravel in the English county of Worcestershire, he could not have imagined that he was about to change both his own career and science's understanding of the distributions of species. Not expecting to do anything other than take a cursory look around, he had not brought any collecting gear with him. But nestling among the mammoth and woolly rhinoceros bones – which he already knew could be found there – in deposits dating from the last ice age, he spotted some dark matter that piqued his curiosity. So Russell rapidly consumed his emergency rations and scraped out the dark material into his biscuit tin. Back in the lab, he inspected it under the microscope, and soon realized that the black material was the wing cases and head capsules of ancient dung beetles which had evidently been feasting on faecal gifts that had been deposited by the mammoths and rhinos some twenty thousand years ago.

Off he set, on a tour of the gravel pits of southern England. The beetle kept reappearing, and he even found 150 different individuals of the same species in a pit near the village of Dorchester-on-Thames in Oxfordshire. Russell was a beetle expert, but he couldn't identify it. The beetle belonged to the genus *Aphodius*, of that he could be sure, but what was it? Eventually, after exploring the musty drawers of museums and consulting with colleagues, he tracked it down. It went by the name of *Aphodius holdereri*. No wonder he didn't know what it was. It is a species known only at altitudes of over 3,000 metres in Tibet.[12] Any contemporary ecologist or evolutionary biologist inspecting the present-day distribution of this insect might have assumed that the species had evolved in isolation, high in the mountains there. But no, the Tibetan population is simply the last refuge of what was once the commonest dung beetle in north-western Europe.

Russell was on a roll. The dark matter revealed dozens of species of beetle, all of which used to live in Britain but which now live thousands of kilometres away, many in the Arctic, in frigid environments similar to those that were widespread in Europe during the last ice age. And there were other bizarre examples. One of these once-British beetle species is now confined to the Pyrenees. Another today lives only in eastern Asia. In many respects, they are just like geladas. Fossil bones that date back to the Pleistocene epoch (a time of intermittent ice ages from 2.6 million to 11,700 years ago) reveal that relatives of geladas formerly roamed from the southernmost parts of the African continent, up through the Congo, across northern Africa, Spain and elsewhere in southern Europe, and eastwards into the Indian subcontinent. Now they face their last stand in the mountains of Ethiopia. The gelada and ice-age dung beetles have behaved in rather similar ways.

Russell went to war with the conventional view that species evolved where you find them today. Starting with a bit of dirt scraped into a biscuit tin, he demonstrated that many species currently live in places where they happen to survive, rather than where they originally evolved. Some species were much more localized in the past, others more widespread; some were not necessarily any more or less wide-spread, but they nonetheless lived in different places. This idea soon took root among scientists who study the history of life on Earth but, surprisingly, its importance is still only partially appreciated by many researchers who are primarily interested in the present-day ecology of the world and by those who attempt to protect it. Dynamism is the norm, not the exception. It is how species survive times when the world's climate changes.

Half a century later, inspired by Coope and by his botanist contem-poraries, I grabbed my spade and set off down the garden. As a child, I may have dreamt of digging a hole through the centre of the Earth to reach Australia but, as a fifty-something academic, I took on a more practical challenge: to dig a hole so that I could visit the last ice age.

The top twenty centimetres of my horse pasture proved to be quite ordinary, sandy soil. This soil is a record of recent history, containing most of the roots of the meadow plants and occasional fragments of brick, glass and pottery from the last three centuries. This hay meadow is biological community number one. I dug on through increasingly sandy soil, which was churned from having been ploughed; it would have supported crops for a thousand or more years. Biological community number two is the cereal field. Before that, forest trees would have grown in place of these crops, in a Holo-cene epoch landscape that existed from five to ten thousand years ago. Biological community number three – deciduous forest. Then I reached a rusty-looking, crunchy layer, a hard pan of iron-rich encrustations separating the browner soil from what is below. Below it was almost pure orange sand, as though I had bought it from a builder's merchant, although streaked by the odd root and the darker brown burrows of earthworms. Biological community number four – this was presumably different again, but I could not see anything that told me what grew there. I kept digging and found more dry sand, then, close to a metre down, I came across clay, reflecting an earlier history. The clay was so thick and hard that it required a much smaller and sharper spade to penetrate it at all. Then, when it rained, it turned into gloopy, adhesive clay and held the water like a pond.

Sand was sitting on top of clay, with the sharpest of possible transitions between the two; it was as if someone had dumped an enormous pile of sand on top of a flat plain of potter's clay. I celebrated, spade in hand. An hour and a half of digging and I had reached the ice age.

Resting in my newly dug hole, I was standing on clay that was laid down over fifteen thousand years ago. At that time, my 'land' was at the bottom of a massive body of water called Lake Humber. Biological community number five – glacial lake. I could imagine myself fishing for red-bellied Arctic char, protected from the cold winds by reindeer-hide clothing, my ears warmed by an Arctic fox hat, its white fur the most insulating in the world. Looking northwards to where the City of York has since been built, I imagine the lake's ice-covered shore, which at that time would have been the bounds of the habitable world. Crumbling ice cliffs melted into Lake Humber's chilly waters, the edge of a white massif that was one of the world's great geographical features of its day. A plateau of ice extended from the western limits of Ireland as one uninterrupted ice sheet over northern Britain, across the North Sea, northern Germany and Poland, up through Scandinavia, and on across the Arctic Ocean, north to the islands of Svalbard.

Summer melt-water fed into Lake Humber. As the ice melted, stones, sand and other coarse materials embedded in the ice were dropped as long ridges at the toe of the glacier, forming the Escrick and York moraines; today, low, gravelly hills that I drive up every morning on my way to work. But the fine minerals – the clays – were free to float on into the middle of the lake. To the south and south-west were the river systems of eastern England, bringing more suspended clays, which also floated out into the lake before settling at its base. There was no escape for these products of erosion, blocked by ice to the north, east and west, and by low hills to the south. The impounded waters spread out into a massive lake. Gradually, Lake Humber filled with clay, silt and sand.

If I take out my microscope, I can find occasional head capsules of water-inhabiting midges in the grey gunge, and signs of long-gone Arctic diatoms, the suspended algae that once formed the base of the food chain and which would have supported the Arctic char and dagger-billed great-northern divers that were hunting fish.[13] Arctic fox and snowy owls would have inhabited Lake Humber's southern shores. None of these species lives in this region today, though divers continue to fly past the Yorkshire coast in winter. Still, it is relatively easy to imagine the world that existed between twenty thousand and fifteen thousand years ago.

So it was until the day of the great Yorkshire flood. As the climate warmed approximately fifteen thousand years ago, the ice dam broke and Lake Humber drained away in a flash into what is now the North Sea,[14] revealing a plain of clays, silts and sand. As frigid and desiccating conditions returned a few millennia later (11,500 to 12,800 years ago),[15] the meagre vegetation cover that could gain a footing across this muddy plain was insufficient to prevent the wind from rearranging the former lake's deposits. Wind-blown sand formed dunes across parts of the lake bed – now the Vale of York and Humberhead levels – and my house and land are perched on top of one of them. It was wind that dumped the enormous pile of sand on the clay-covered plain. The marshes of the Vale of York and these ancient dunes, which would have been covered by steppe-like vegetation, were home to furry lemmings, honking flocks of wild geese, antlered red deer, bone-crunching spotted hyenas, ancestral bison, woolly rhinoceros with their sweeping horns and ivory-bearing mammoths: a perfect hunting ground for summering ice-age humans.[16] Biological community number four was a mammoth steppe. Most of the animals and plants that lived on that ancient steppe are today citizens of the high Arctic, although spotted hyenas, like geladas, went in the opposite direction. They survive only in Africa. The rhinoceros and mammoth were consumed by our ancestors.

Coope was absolutely right. These ancient communities of animals and plants reveal that the biological world has been turned on its head repeatedly since the end of the last ice age. Lake species were replaced by dune, steppe and wetland species. These were supplanted by forest species that were in turn usurped, as humans converted the land, by species of open farmland and field margins. Then I made the land into hay pasture. And today, new arrivals are turning up as the climate warms once more, changing the biological community again.

The moment we focus on timescales that exceed ten thousand years,

Horse pasture;
last ~40 years

Forest then farmland;
last ~10,000 years

Iron-rich hard-pan;
last ~10,000 years

Ice-age mammoth steppe and dunes;
~11,500 to 12,800 years ago

Ice-age lake clays;
~15,000 to 20,000 years ago

The changing soils of the last twenty thousand years, in the Vale of York in England. At least five completely different sets of animals and plants have lived here during this extremely short space of time ('Arctic' lake, mammoth steppe, forest, cropland, pasture), demonstrating the dynamism of the biological world. Human-caused changes to the climate are now allowing a sixth set of species to become established.

we come to realize that species move around the surface of the planet when the environment changes. The set of species present on my own bit of land has been almost completely replaced *at least* four times in the last fifteen thousand years, and *at least* forty times in the last million years of repeated climatic changes. Biological communities are transient. Rearrangement followed by rearrangement is the norm. That is how species survive climate change. They move around. This is true even of species that are today confined to shrinking distributions in isolated mountains. They have survived because they have been able to move, even if that has only been up and down the same mountain. The greatest risk they face in the future is the possibility that they may not be able to move again.

Standing in the squidgy hole I have just dug in my garden, I begin to appreciate something quite fundamental about life on our planet. Not only have the inhabitants of my small spot on Earth *always* been interlopers, but this is also true of every spot on Earth. Species come and go with the vagaries of the Earth's climate, and with any other great environmental change. In due course, today's interlopers will be replaced by the next set of temporary residents. Any attempt by humans to keep things just as they are is utterly pointless.

Wherever you live, it is worth taking out your spade. Not everyone is lucky enough to be living on top of geological deposits that were laid down during the last ice age and, even if you do, town planners may take a dim view when you start digging up the local park. But at least take out a virtual spade and imagine the history of your home.

Twenty thousand years ago, Canada and large parts of northern Europe were under ice, so all the species that live in these areas at present have arrived since that time. Ice sheets locked up so much of the world's water that sea levels were about 130 metres lower,[17] hence the coral reefs that form the atolls of the Pacific and Indian Oceans have grown since the ice melted. All the species that live at the intertidal boundary between the land and the sea have moved to today's coastlines since the climate warmed and the ice melted. At the height of the last ice age, many North American broad-leaved trees were living on now-submerged continental shelves under the waves of today's Caribbean Sea – the animals and plants that live in North

America's great eastern forests are nearly all recent arrivals. The Amazonian forest contained mixtures of Andean, Amazonian and dry forest trees not seen today,[18] the trees having since gone their separate ways. There were enormous lakes in the now-arid Great Basin of western North America, and the Sahara Desert was covered in trees and lakes as recently as 10,500 to 5,500 years ago.[19]

The species that are found wherever you happen to live were different – often radically different – from what you see today. When the climate warmed after the last ice age, the distribution of species started to shift, dramatically rearranging themselves over the ensuing thousands of years. Boreal willow grouse and Arctic ptarmigan, which turn white in winter as camouflage against the snow, were hunted by our ice-age ancestors across the lowlands of central Europe, in locations which these birds no longer inhabit; they subsequently spread into the high mountains and to the far north of the continent, where they live today.[20] Greenland collared lemmings did not live in Greenland but were found in a band across the United States, far to the south of their present range. This is perfectly normal. The redistribution of species in the last twenty thousand years has simply been the last act of the Pleistocene epoch, which has experienced a roller-coaster of alternating extreme cold (ice age) and warm (similar to now) interludes for the last million years. The ancestries of all species stretch across this period of instability. Individual species have survived by moving, and every location has seen a kaleidoscope of changing biological communities as new species arrive and others depart.

It is hard for durations of longer than ten thousand years to grab our attention. Even so, it is essential that we learn lessons from the past. The magnitude of human-caused change to the climate between the middle of the twentieth century and the end of the twenty-first century will be on a par with changes that normally take tens of thousands of years. Moreover, the new world climate is likely to become warmer than it has been for three million years. If we wish to understand how the biological world responds to and survives climatic shifts, it is these longer timescales to which we must pay attention. Today, we regard these past transformations as natural, and we accept without question that species are found where they should be. We think of the new vegetation and reefs that developed at the end of the last ice age, and of Greenland lemmings living in Greenland, as the way that the Earth is 'meant to be'. However, while there are plenty of reasons why we would wish to reduce the rate at which humans alter our planet's climate, there is no logic in defining this past change as good and natural and at the same time describing more recent and future change to the distributions of species as regrettable and unnatural. This is to impose an inappropriate human sense of territory and duration on the unique histories of every population of every species. Looking back from the twenty-first century to the last ice age, we appreciate that our ice-age selves would have been wholly mistaken to have resisted changes to the distributions of species that took place as a consequence of the climatic warming at that time.

Whatever period we are considering, species have moved to take advantage of new opportunities that have arisen from time to time, just as they died out in places where conditions became unsuitable. It is the way our biological planet works. It is the same again today.

The burnt ochre and chocolate-coloured comma butterfly, flexing its ragged-edged wings on the fissured trunk of my apple tree near York, contemplates its first flight of spring. It is a biological beneficiary of humanity, spreading northwards as the climate has warmed. Likewise, the red mason bee sunning itself on my human-built wall has been aided by our existence: it has been spending most of the day visiting spring flowers that I planted, constructing pollen-filled cells for its offspring in the cracks of my crumbling mortar, and enjoying a warmer climate.

No specific butterfly knowledge is required to work out that the comma butterfly has spread northwards, nor is there any need to spend hour after hour searching through naphthalene-drenched museum drawers of preserved specimens that have seen better days – much as this activity brings joy to entomologists. The UK Biological Records Centre and the charity Butterfly Conservation have already amassed as many museum records as they can find, inspected the old literature and collated the records of amateur butterfly spotters. These they have brought together as one electronic database, which can be examined by anyone.[21] The records show that the comma butterfly started to appear in Yorkshire in northern England during the

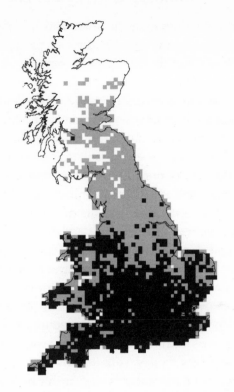

The comma butterfly has spread approximately 350 kilometres northwards since the 1970s because humans have warmed the climate by around 1°C, enabling it to live in regions that were previously too cold (black squares represent records between 1970 and 1982; grey squares show new locations that were colonized between 1983 and 2015. The side of each square is 10 kilometres).

1970s, which coincided with the first major survey of British butterflies. This survey allowed John Heath, Ernie Pollard and my brother,
Jeremy – the little boy in the wedding photograph in Chapter 2 – to
publish the first national butterfly atlas for Britain in 1984. At that
time, Yorkshire was as far north as commas could be found.[22] However, the butterfly was soon to continue its path towards the North
Pole, reaching Aberdeen in north-eastern Scotland a mere twenty-five
years later.

Not only have we provided the comma with a suitable climate, we
have also provisioned it with food. The comma's caterpillars feed on
wild hops in my hedgerow, a plant that grows there because it escaped
from the hop gardens that were planted in this region to supply the
brewers of Yorkshire. These included Webster's Brewery, founded by
Samuel Webster in 1838, my own great-great grandfather, whose portrait looks down on me as I write these words. But for the brewers
who needed hops, and the farmers who planted the hedgerows as well
as the crop, the butterfly's caterpillars would not have had any hops
to eat. Thanks to Victorian naturalists, we know that commas did
indeed fly across the Vale of York in this heyday of hop growing, but
then they retreated southwards, before recolonizing in the 1970s.
Hops are not that common any longer in this part of the world, since
Yorkshire brewers source their ingredients from further afield. But
stinging nettles abound, beneficiaries of farmers spreading liberal
amounts of nitrogen and phosphorus fertilizers across the landscape
to enrich their fields. Nutrient-loving nettles are thriving – a success
story of their own – providing yet another opportunity for the comma
butterfly.

My friend and colleague Jane Hill, whose own gleaming ginger
pelt gives the mason bee a run for its money, wondered why the
comma had been quite so spectacularly successful in recent years, so
she and postdoctoral researcher Brigitte Braschler decided to investigate. They set off in search of adult butterflies that had just woken up
from their long winter snooze. During the dark, cold winter months,
these butterflies sit motionless, the blackened, ragged undersides of
their wings virtually indistinguishable from dead winter leaves. As
spring arrives, they start to appear, warm themselves up, seek nectar
to drink, find a mate and, in the case of the females, begin looking for

any suitable plants where they will be able to deposit their eggs. Jane and Brigitte caught some female butterflies from near York, and others from further south, and brought them back to their laboratory, where they kept the butterflies in a toasty-warm room to lay their eggs.

As the eggs hatched, some of the caterpillars were given hops to eat, the plant that British comma caterpillars traditionally ate, whereas others were provided with a delectable diet of stinging nettles. After months of tending their lepidopteran family, Jane and Brigitte found out that the caterpillars from the newly colonized region of York survived better on a diet of nettles than on hops, compared to the caterpillars whose parents had been caught in the south of England. Moreover, this advantage was enhanced at higher temperatures.[23] It turned out that the northern commas, which had established a new population in the Vale of York and then continued to spread even further north, have recently evolved a liking for stinging nettles under a human-warmed climate. This, combined with the burgeoning population of fertilizer-dependent nettles, has been propelling the butterflies northwards.

Of course, the butterfly itself has no other purpose than to mate and lay its eggs. It is itself oblivious to the changing world. It just happens. Each individual lives and dies, and moves a bit, and breeds. And on it goes. Over the generations, climatic conditions became more favourable and small changes have seemingly taken place in the butterfly's ancestors' genetic code that help it eat nettles; changes that have been sufficient for the comma butterfly to have moved the 350 kilometres from York to Aberdeen in just twenty-five to thirty years. It will soon reach the Arctic Ocean. It is easy to see that fairly modest changes to the distributions of each species that we can observe on a timescale of a few decades are sufficient, if replicated across millions of species, to rearrange life on Earth over the thousands of years of human influence. All it takes is a little time, just as it did when species were moving around the world as they responded to climatic changes during the Pleistocene epoch.

The climatic changes that humans have already wrought have brought a cavalcade of animal adventurers to new locations. A few months after spotting the comma and mason bee sunning themselves, I accompanied my domestic wolf for a spin around the garden and decided to count the butterflies. There was a rusty comma, perhaps the offspring of one that I had seen in the spring, this time seeking out the nectar of a bramble flower. Incoming species number one. Female ringlet butterflies, which have perfect white rings marking their deep-brown undersides, were plying up and down, dropping their eggs into tussocky grasses at the edge of the field. Two. Male speckled woods sat perkily in shafts of sunlight between the trees, flying out to intercept females and chase off rival males. Three. Gatekeepers were flitting along the hedgerow and feasting on marjoram flowers on my rockery. Four. A deep orange Essex skipper was darting back and forth between purple knapweed flowers – a butterfly whose name is associated with the more southerly county in England, whence it came. All five were absent from this land a mere fifty years ago, before they expanded their ranges when the climate warmed. Five out of thirteen different species I saw that day – over a third of the species flying in my garden – were there only because of human-made changes to the climate.

These were not the only five to benefit. When I moved into my present house in the year 2000, I saw one sulphur-yellow brimstone in the entire year, whereas a decade and a half later I was able to count half a dozen soaring over the buckthorn bushes in just five minutes. Equally, small skippers – which can be distinguished from the Essex skippers only by the colour of their antennae – would have been much rarer fifty years ago. If we count the brimstone and small skipper as well, then half the species flying in my garden on a summer day in 2016 were beneficiaries of climate change. And it is not just the butterflies. The red mason bee has been joined by tree bumblebees, colonizing from the south in the last few years, and I spotted a little egret earlier in the season. The egret is a gleaming-white heron, identified by its brilliant-yellow feet, which trail behind as it flies along drainage ditches searching for frogs and insect larvae. This bird bred on the south coast of England for the first time in 1996, and now I can see it from home – it has colonized more than half the country in twenty years. Humans have changed the climate, and the distributions of species have changed as a result.

An inexorable march of the world's wildlife is under way. Moths deep in the forests of Mount Kinabalu in northern Borneo and trees growing on the forested slopes of the Andes have moved to higher altitudes as the climate has warmed. Amphibians and reptiles now live at higher elevations in the mountains of Madagascar. Birds have shifted higher in New Guinea, and they have moved up from the lowlands in Costa Rica. Mammals and birds have moved to higher elevations in the Sierra Nevada in California. Birds have expanded their ranges northwards across the European and North American continents. Plants in European mountain ranges have shifted upwards. Fish, marine plankton and shellfish are steaming northwards in the seas and oceans of the northern hemisphere, and heading towards Antarctica in the south. Warm-water Australian fish have colonized Tasmanian reefs that used to be too chilly. Even those animals that are unaffected by temperature and rainfall directly are living in places where the vegetation they eat and the animals that prey on them have already changed. The footprint of human-caused climate change is ubiquitous.

This is a story of gain as well as loss. Around two-thirds of the species that researchers have studied in recent decades have shifted their distributions in response to climate change, becoming commoner in those places where the climate has 'improved' for them.[24] Animals are moving towards the poles at around 17 kilometres a decade.[25] This is the equivalent of picking up every individual of every animal species and moving it just over four and a half metres every day. This has been going on every day of every year since human-caused climate warming became evident in the mid-1970s. Of course, this is not quite how it happens. In reality, a butterfly occasionally flies an unusually long distance, perhaps five kilometres, and establishes a new colony further north – or south in the southern hemisphere. A few generations later, the new colony has grown large enough that individuals leave it and set off on the next leg of their journey towards the poles. Likewise, a seed may be blown a long distance by a storm, travel inside the stomach of a bird, or be moved in the mud attached to the tyre of a car. The journey is a series of steps, not a smooth progression across the world. But averaged out over a few decades, the overall consequence of all these steps is the progressive movement of vast numbers of species, as they die out in places that are now too hot or dry and colonize new locations that have recently become warm enough for them to breed. Keep this going for a few centuries and we have a new biological world order. If my garden butterflies are anything to go by, we are already a third of the way there.

The ubiquity of this movement across the surface of the Earth means that something like two-thirds of animal species are already living in at least some 'new places' where they could not have survived as recently as fifty years ago, while they have disappeared from other locations. As the present century unfolds, the overlap will progressively decrease between the 'historical' distribution of each species and where that species will then be living. As we found when we contemplated the ice ages, the idea that the distribution of a species is fixed is outmoded. It can still be convenient to think of species as being native to South America or to Eurasia, for example, but it is increasingly unrealistic to suppose that there is anything special about the precise locations where humans first documented the presence of a given plant or animal. It was one frame in the passage of time and the movement of species, of no more and no less significance than any other frame.

As the climate warms, these new opportunities will increase. The average temperature of the world is only about 14°C, whereas most species reach their peak physiological performance well above this. Many tropical and subtropical plants are damaged below 10°C to 15°C, so they have the potential to spread when temperatures increase. Planktonic algae commonly reach peak performance at 15°C to 30°C, so they are likely to spread as the world's oceans warm.[26] The average annual temperature is about 10°C in cool Britannia,[27] yet most British insects become fully active only when temperatures are in the upper teens, and they generally do best above 20°C. On the whole, more species like it hot than cold.[28] This means that the 1°C global warming that has taken place between the late 1800s and the 2010s has unleashed large numbers of heat-loving species. Tropical species can start to spill out into the subtropics, subtropical species colonize temperate regions, and the inhabitants of the temperate zone can try their luck in the polar world; and this has the potential to increase the diversity of those places where they arrive.

This is already happening. Many immigrants from the south have arrived in my garden in recent times, but I am not aware of any northerners that have disappeared. If we consider this more systematically, the diversity of butterflies has increased in most parts of Britain as a consequence of southern species expanding their distributions.[29] It is similar elsewhere. The number of low-elevation bird species that moved uphill into Costa Rica's Monteverde cloud forest between 1979 and 1998 greatly exceeded the number of higher-elevation birds that died out:[30] golden-crowned warblers, lesser greenlets, flamboyant keel-billed toucans and twelve additional bird species from the lower slopes began to nest high up the mountain, alongside the emerald-green and scarlet resplendent quetzals that characterize Central America's dank cloud forests. More species arrived than disappeared, increasing the number of bird species in the forest. If we consider the entire world's land surface, climate change is liable to increase plant diversity across much of it.[31] Given how many species live in the hottest parts of the world, it is not surprising that the average biological diversity per square kilometre of the world goes up when the climate warms.

There are three important caveats. First, we should not forget that the Ethiopian wolf, Kinabalu moths and other species are endangered. This will reduce the total number of species on Earth in coming centuries – although geological history tells us that additional species are likely to evolve if the Earth remains hot for a million or more years.[32] The second is that species need to be able to reach the new locations where they will be able to thrive. Plant diversity will not increase unless seeds are able to get to the places where they can grow – most likely, this transport will be aided by us. The third caveat is that diversity goes down where there is not enough water. This is the real concern. Biological diversity thrives where it is hot, but it does so only if there is enough moisture for plants to grow, and for plant-eating insects to digest them. When my former colleagues Rob Wilson, David Gutiérrez and Javier Gutiérrez hiked around the increasingly hot and dry mountains that surround Madrid in central

Spain in the 2000s, they discovered that the average number of butterfly species was lower than it had been in the early 1970s, before the climate had warmed.[33] This is because the relationship between temperature and diversity is reversed there. Go butterfly hunting in the Sahara Desert, and you will find fewer species than in the cooler Spanish mountains – the Sahara is too dry. If increasingly Sahara-like conditions start to spread, the mountain butterflies will disappear, and there are not so many drought-adapted species available to replace them. Thus, climate change brings reductions in diversity in some places, just as it generates increases in others. Averaged across the whole world, however, the amount of rainfall is increasing. More water evaporates from hotter oceans, and it comes back down again as more rainfall. Places that become warmer and remain wet usually gain biological diversity, as do dry places that experience increases in rainfall.[34]

This is reasonably good news, but it is not an argument advocating that we stop worrying about climate change. A three-quarters-full cup is still a quarter empty. Coral reefs are threatened by hotter and more acidic water and, if we want to keep Ethiopian wolves alive in the mountains of Ethiopia, we must minimize the greenhouse gases that are emitted into the atmosphere. Nonetheless, the climate is changing, and the basic expectation of a warmer and slightly wetter world is that the diversity of many – and perhaps most – regions in the world will increase. This is what we are starting to see, accomplished by the movement of species across the surface of the Earth. We need to accept and even encourage this movement because botanical and zoological world travellers will form the basis of the world's new ecosystems, just as they have when the climate has changed in the past. These travellers are the future of life on a warmer Earth.

5

Pangea reunited

The exuberant green growth of palm seedlings spilled out of the forest and on to the edge of the winding road. The deeply dissected leaves of Chusan windmill palms flickered in my headlights. This palm grows further north than any other in the world and, at last, I was seeing it for myself, in the wild. I had tried to grow it back home in Yorkshire, but failed. The seeds I planted germinated well enough, but the seedlings had perished, each meagre bout of summer growth cut back by the following winter's cold. In the slightly milder climate I was now visiting there was hardly enough space to accommodate the chest-high, trembling fronds that formed an evergreen thicket. I swerved closer for a better look. Too close. The high stone kerb penetrated the side of the tyre and immobilized the hire-car that I had picked up earlier that same day. And there I was, stranded in a foreign country, surrounded by a living herbarium of exotic greenery, unable to speak the local language and in a vehicle without sufficient tools to replace the wheel. I was left attempting to communicate in sign language with some locals whose gas station doubled up as the village bar, rather than as a workshop.

From the surrounding vegetation full of Chinese palms and Nepalese camphor trees, I could have imagined that I would need to speak Mandarin or Nepali, hence my inability to communicate might have seemed understandable. But I was stuck by the shore of Lake Maggiore in Switzerland, and only Italian would suffice. Our educations had failed us all. We gesticulated with enthusiasm and attempted to communicate in pidgin French, German and Spanish. We phoned Giovanni's wife, who did speak a few words of English. Several hours later, I continued on my way, the kind clientele of the bar-cum-gas-station having persuaded an Italian vehicle rescue service to cross the border into Switzerland and make running repairs to my car.

The following day revealed a flickering, Hockneyesque palm-fringed hotel swimming pool with a backdrop of snow-capped peaks and forested mountainsides, accompanied by a soundtrack of cheeping Italian sparrows. One could, for a moment, have believed man and nature to be in harmony, yet Lake Maggiore is a postcard from the Anthropocene. The shores of this enormous alpine lake have been altered by humans ever since the ice-gouged trench of the retreating glacier filled with water. Ancient grasslands and forests waxed and waned with the vagaries of the climate at the end of the last ice age,[1] yet all these changes already bore the signature of humanity. Our ancestors had killed off Europe's hippopotamus, the forest elephants and rhinoceros, as well as the lions and Etruscan bears. As a consequence, this new megafauna-free forest was much darker than it would otherwise have been.[2] The Stone Age around Maggiore gave way to the Copper Age, about seven thousand years ago, then the Bronze. The forest was cut and cut again: for fuel to cook food and smelt metal, and to build houses and livestock pens. Parts of the forest were cleared to cultivate crop plants that had originated in the Middle East, and domestic animals arrived. The period of humans deliberately importing useful new species had begun. The Romans introduced sweet-chestnut trees for their seeds. They carefully peeled back the prickly rind of their fruits to reveal the glistening nuts, which they ground into flour. By AD450 the chestnut had become established as a major part of the forest in the hills above Ancona, near to the head of the lake – a cultivated species gone wild. The landscape never stood still. Nor did the species that lived in it.

The speed of transformation accelerated with the initiation of the Industrial Revolution. In 1808, the small lakeside town of Intra became home to Italy's first steam loom, a picturesque 'Manchester of Italy'. Industrial wealth meant rich people, and rich people in the nineteenth century craved country villas and gardens. Their enthusiasm for exotic plants to embellish the land became a passion and, to this day, the shores of Lake Maggiore are dotted with plant nurseries and garden centres. A new era had dawned. From a modest start, in which humans had deliberately transported a few different kinds of crops, livestock and medicinal plants over thousands of years, the importation of new species had gone into overdrive. The age of European palm forests was about to commence.

As I set off the morning after my nocturnal adventure, the early-spring mountainside above Locarno looked surprisingly green. On closer inspection, deciduous trees that had not yet spread their bright summer livery could be seen standing as twiggy sentinels above the billowing sombre green crowns of the shorter evergreens. As I entered the forest, the insanity of it struck me, and my meagre botanical skills were found wanting.

Some, though, were already familiar to me, especially the leafless species that would have lived in the original forest. The deciduous oaks, linden trees, ash, alder and Roman-introduced sweet chestnuts, all of which drop their leaves in the winter, also grow at home in England. The evergreen holly and ivy were familiar too. Prickly holly trees punctuated the under-storey, and native ivy clambered over the trunks of deciduous trees, clothing their brown stems in green and scrambling across the forest floor. There were also species from elsewhere in Europe, the aromas of which were more familiar to me than the plants themselves. Cherry laurels could be identified by the cyanide smell of their crushed leaves. They were native to south-eastern Europe and the Black Sea but now thrive in Maggiore's lakeside forests. Bay laurel, familiar to cooks for its essential oils, was prized as a symbol of high status in the Greek and Roman worlds, and a laurel wreath was worn by the ancient god Apollo. This tree grew elsewhere in the Mediterranean in the past, but it has been planted more widely and now regenerates with abandon in these new forests.

Others were less familiar. Nepalese camphor trees were growing high into the forest canopy, equally identifiable by their odour. Transported from the Himalayas, camphor is today a tree of the Alpine foothills, part of the new Insubria forest (Insubria is the region that spans northern Italy and southern Switzerland; Lake Maggiore straddles the two countries). There they grow, alongside all manner of Asian plants: suckering bamboos, silverthorn, broad-leaved privet and box-leaved honeysuckle, willow-leaved cotoneaster from western China, loquat with their bright orange fruits from southern China,

Henry's honeysuckle, the rampant evergreen Japanese honeysuckle whose vanilla-scented yellow and white flowers give rise to small black berries, and sweet-scented Japanese mock-orange and camellia, whose waxy-pink flowers are visited by European bumblebees. Not to be outdone by floral beauties from the Orient, occasional creamy-flowered bull bay magnolias from the US states of Virginia and Florida can be spotted, as well as Oregon-grapes, and silver wattle from Australia. It is a diverse and fascinating forest, however unlikely its origins. Meanwhile, there seems to be no real threat to the European deciduous forest trees. Higher up the mountain, it is still far too cold for the usurpers to survive. The entire forest that encompasses the shores of Lake Maggiore and the chilly upper slopes is far more biologically diverse than it used to be.

While industrialists were responsible for transporting these plants from far-flung parts of the world, their journey from garden to forest was accomplished by a team of European and Asian seed-movers and seed-buriers. The nutritious windmill palm fruits seem irresistible to European blackbirds, whose glistening black feathers and gaudy plastic-orange beaks adorn grassy lawns throughout Europe. Even they are not quite what they used to be. Originating as shy, forest-dwelling birds that shunned human presence, it is now hard to find a European town or city where they do not hop across lawns and sing from the shrubberies.[3] Maggiore's waterfront towns and villages abound with these newly confident birds, as well as feral rock doves and collared doves, the latter having spread from India and into western Europe in the twentieth century. Yet more success stories. Then there are jays and wood pigeons, with their own urbanization stories, and Italian sparrows, the new species formed by hybridization between Asian house sparrows and Europe's own Spanish sparrows. Add a host of mice and squirrels, which run off with seeds and bury them in caches to secure themselves a food supply for the winter, and the full company of seed-movers is assembled.

This team transports seeds in astonishing numbers. Seedlings sprout along the bases of hedgerows, in the back of every garden border, and under any tree where a defecating or regurgitating bird might sit. A profusion of saplings can be found in the adjacent forest. Wherever the wildling palms have grown large enough to produce their own fruits, the under-storey is a veritable jungle of second-generation fronds. A few seeds have been transported half a kilometre, or even further, from the nearest palm-containing garden, which is about as far as most of the birds can be expected to fly in one trip. Within a palm generation or two – a human lifetime – virtually all of the warm, forested slopes that line the banks of the lake will be verdant palmeries.

Careful examination of garden records from nurseries and from the large estates of the new industrialists indicate that most of these gone-wild plants were cultivated in the region's gardens for over a century before they started to grow in the new Insubria forest. The giant forest trees that clothe the lake shores are deciduous trees. Their autumnal golds and yellows signal that they will drop their leaves in preparation for the cold of winter, when blasts of freezing air descend towards the lake from the surrounding High Alps. The undergrowth of hazel, blackthorn, hawthorns and dogwoods follows suit, although European privet and rambling brambles hang on to some of their leaves if the winter is not too harsh. Ivy scrambling across the forest floor and a few holly bushes were once rare spots of greenery in the winter-brown forest. And so it might have stayed, with hundreds, or even thousands, of evergreen plant species carefully nurtured in lakeside gardens, protected from damaging frosts by a cadre of attentive gardeners, while the deciduous trees continued to dominate the mountain forests. However, the industrial moguls who had acted as plant hauliers from the east also brought us global warming, and the severity of winter frosts abated in this region during the second half of the twentieth century.[4] With warmer winters, the evergreens gained an advantage. Palms and camphor could now grow more luxuriantly than linden trees. All that was required was for the blackbirds and squirrels to assist their escape from the industrialists' gardens. The transport of plants across the world, climate warming and semi-tame animals all combined to bring about the transformation.

It is not just the forest. The lake itself is full of foreign species, most of which have been introduced in the last century. The pike-like zander fish is an import, although only from elsewhere in Europe. An oversized specimen of this denizen of the deep had to be tracked down by police harpoonists after a troublesome individual bit chunks

out of six bathers – in compensation, its walleye-flavoured meat was served to the tourists. One hopes that the fish had had time to digest its last meal before their feast. Most of the additions, however, have originated from North America. Largemouth bass, which is so feisty and prized a catch that it is an official state fish in Mississippi, Alabama, Georgia, Tennessee and Florida, was introduced as early as 1930. The enthusiasm of anglers has provided this fish with a ticket to the world. The pumpkinseed sunfish is also American. It wears impressive black-and-orange 'false eye' markings on its sides to give large predatory fish and diving birds the impression that their quarry is far more dangerous than it is. The American bullhead catfish has a

The lower slopes surrounding Lake Maggiore in Switzerland (above) are home to exotic gardens, where windmill palms are grown. The palm fruits are consumed by European blackbirds (opposite, top), which deposit many of the seeds in the nearby forest, forming a dense palm under-storey, through which it is possible to see the bare stems of the original forest trees (opposite, middle). Palm crowns, Nepalese camphor trees and other Asian and North American evergreens now join the deciduous trees to form a forest unlike any other that exists (opposite, bottom).

different strategy: it keeps out of sight, hoovering up anything from plant debris to insects and rotting animal matter in the murkier parts of the lake, escaping notice by feeding at night. Altogether, over a third of the thirty-two types of fish in Lake Maggiore are introduced species without, as far as is known, any 'native' species becoming extinct as a consequence.[5] So the lake contains many more kinds of fish than before humans appeared on the scene, just as the forest now supports newly arrived populations of exotic trees and shrubs.

And so Maggiore continues its journey into the Anthropocene epoch – an ecological and evolutionary melting pot of the world's species. European and Asian birds and mammals mix with African-origin humans. Together, these animals are moving the seeds of North American, Asian, South American, European and Australasian plants into a forest of what used to be European species. It is an international blend, rich in diversity. It is a novel ecosystem or, in the parlance of some ecologists, an Anthrome.[6] It is a human-altered land.

The success of Asian and North American trees is partly down to the absence of similar European evergreens, which died out during the ice ages. Geologists and botanists have spent the last century staring down their microscopes at the rot-resistant walls of pollen grains, preserved leaves, and at the internal structure of fossilized tree trunks, which they have unearthed from the bottoms of lakes and the alluvial outwash of rivers that flowed three million years ago. In so doing, they have discovered the remains of large numbers of extinct trees that used to live in Europe, including magnolias, which still survive in eastern Asia and in North America. Jens-Christian Svenning, a tall, crazy-golf-playing scientist from Aarhus University in Denmark, worked out that as many as thirty-one genera[7] of trees that were native to Europe between 5.3 and 2.6 million years ago have since become extinct, whereas thirty-five have survived in the region.[8] If you had taken a grand tour of Europe 3 million years ago, you would have encountered double the diversity of native trees.

This rich European forest became impoverished by a succession of ice ages, which eliminated nearly all the cold-sensitive trees (those unable to persist in places where the average annual temperature is below 0°C), including most of the broad-leaved evergreens. In contrast, cold-sensitive trees and shrubs did manage to survive the ice ages in more amenable climates along the Caribbean fringe of North America and in south-eastern China. Now, globe-trotting humans are bringing them back. Relatives of the original European trees are coming 'home' during a period when the human-warmed European climate increasingly resembles the conditions that existed 3 million years ago. Although the returnees are not quite the same as the original species, they can nonetheless be thought of as nearly native trees that are flourishing in Europe once more. The rediversification of European forests is under way.

Conservationists and many ecologists are not happy about this. A hatred of foreign species is regarded as perfectly acceptable, and to think otherwise is tantamount to heresy. For instance, New Zealand scientist James Russell and British academic Tim Blackburn liken people who do not condemn introduced species in sufficiently strident terms to those who undermine the scientific consensus on *'the risks of tobacco smoking or immunisation, the causes of AIDS or climate change, [and] evidence for evolution'*.[9] And this neophobia has been translated into practical action. At a political level, our governments invest in keeping foreign species out through customs controls, and nearly all countries are signatories to the international Convention on Biological Diversity. As such, taxpayers are committed to the costly control or eradication of priority (aka successful) alien species. On the ground, conservation volunteers are poised to kill invaders when they arrive, and take action to reduce their numbers. For example, Locarno locals are chopping down small patches of evergreen forest on the slopes above Lake Maggiore and planting deciduous linden trees in their stead, in a vain attempt to rewind an unrewindable history. However, we need to ask whether we are just responding to a nostalgic impulse that the world should be as it once was. Today's Insubria forest certainly differs from all previous forests that have existed, but it is still doing all the things that forests do. It is growing, providing wood, nectar, pollen, fruits and seeds that animals eat and transport. It gives shelter, and the roots of the forest trees and lianas stabilize the ground. The forest delivers benefits that humans prize.

How long will it be before the environmental police force of ecologists and conservationists is prepared to step back and decriminalize introduced species that have had the temerity to be successful? Back home in Britain, members of the eco-constabulary are unanimous in accepting as 'native' any species whose ancestors arrived between about five thousand to fifteen thousand years ago. Hence, they are deemed innocent of ecological harm, however common they might be and however large their impacts on other species. Environmentalists are equally of one mind in their condemnation of the gloriously purple rhododendron bush from southern Europe, which was first planted in Britain in 1763. They also love to hate a kind of annual balsam, policeman's helmet, first grown in 1839 and today popping its explosive seed-pods into my roadside ditch. Much loved by bees, the policeman's helmet balsam has such startling pink flowers that it has acquired a remarkable diversity of popular names – ornamental jewelweed, to reflect its beauty; Himalayan balsam and kiss-me-on-the-mountain, after its geographic origin; and gnome's hat-stand, Bobby tops, and Copper tops – as well as policeman's helmet – in recognition of its domed helmet-like flowers. Hardly a historical drama or documentary is broadcast without it putting in an appearance, as the camera crew line up the most attractive flowers they can find to complement their favourite country house. Ecologists of Russell and Blackburn's ilk apparently know better. Himalayan balsam is an invasive foreign species. A couple of centuries is too short a time for an introduced species to be accepted.

The running hares that nibble my meadow and the aphid-sticky sycamore tree in the hedgerow are, however, more of a challenge for the ecological jury; it is unable to decide whether these species belong to a rose-tinted past or are modern interlopers ruining our native ecosystems. The brown hare was most likely introduced by the Romans some two thousand years ago and has been added to the British list of protected species; so it is officially accepted and treated as if it is native. Roman-introduced sweet chestnuts that have been in situ for two thousand years are apparently acceptable around the shores of Maggiore, too, so two thousand years seems to be enough. The sycamore tree was added to the British flora only about five hundred years ago, in contrast, and many conservationists continue to frown upon it (but bow to the inevitable and rarely dig it up). It seems to take somewhere between five hundred and two thousand years to convert xenophobia into love. Any specific duration is hard to justify.

The ancestors of all species that are alive today have flowed back and forth across the globe for many millions of years, as we saw in the last chapter. We should never assume that where we see a species today is where that creature's ancestors originated. Modern humans resided in Africa two hundred thousand years ago and subsequently spread across the world, yet most of us alive today think of ourselves as natives of the regions where we were born, rather than of Africa. When we trace our ancestry, we often focus our attention a few generations back, seeking a sense of place and personal identity. But these past places were also transient locations, and each ancestor transient, as our genes have moved around the planet's surface. It was ever thus. It is completely illogical, then, to hate a fellow human, or another animal or plant, simply because they or their ancestors were somewhere else at a particular time. The location of those genes in one specific timeframe has no special meaning in the history of life.

Humans have not invented the transfer of species between distant locations, but we have dramatically increased the rate at which these events take place. The history of land-dwelling species has been strongly influenced by the movement and locations of the world's continents, which float like oil-on-water islands over the denser planetary mantle and ocean floor. The Earth's continental landmasses of granites and sedimentary and metamorphic rocks weigh around 2.7 metric tonnes per cubic metre, the basaltic ocean crust 2.9, and the Earth's denser mantle is a hefty 3.3 metric tonnes per cubic metre. Gravity dictates that the 'light' continents float. And they travel at a few centimetres a year, about the rate at which our nails grow, inexorably transporting their cargoes of animals and plants across the planet's surface.

Despite the desperately slow progression of plate tectonics, this movement has been sufficient to transport the continents over considerable distances, sometimes bumping together to form one great blob of continental crust adrift on the sea. This was the case a billion years ago, when the land came together into the supercontinent Rodinia, and again between 300 and 175 million years ago, when the continental crust formed Pangea (meaning 'all land'). By then the land had already been colonized by plants and insects, and Pangea was the supercontinent where reptiles diversified and mammals evolved. Almost all the world's land biota lived on a single continent at that time, save island species, most of which subsequently disappeared without trace. Then Pangea split, first into two supercontinents, a northern Laurasia and southern Gondwana, and subsequently into the smaller fragments that became increasingly recognizable as today's continents. But the bits have started to coalesce. The former island that was India rammed into Asia about 50 million years ago, forming the Himalayas in the crumple zone, and South America (previously part of Gondwana) joined forces with North America (part of Laurasia) a mere 5 million years ago.

'Soon' the Alps will resemble the Himalayas and the Mediterranean is likely to be squidged out of existence as Africa continues to steam into Europe. Meanwhile, Australia and New Guinea can be expected to ram into the corner of Southeast Asia, bearing kangaroo gifts. Projecting forwards, some suggest that our presently separated continents will again reunite in a few hundred million years, forming a new great supercontinent. Gradual reconnection of the world's biota a hundred million or more years from now might have been the fate of life on Earth, had it not been accelerated by an unusual ape that evolved in Africa. Ocean-going container ships move species 10 billion times faster than migrating continents, aeroplanes 200 billion times faster. We have set about reconnecting the continents on a much faster schedule.

This is not to suggest that distant locations were totally isolated before humans came on the scene. Many microbes that are critical to the operation of every ecosystem have near-global distributions, blown as dust or attached to the muddy feet of migrating birds. All the ancestors of animals and plants that today live on volcanic islands which emerged from the oceans must have rafted or flown there at some stage. Tortoise ancestors of the Galapagos and Aldabra giants did not originate on those islands. They floated, hitching a lift on ancient tree trunks or mats of vegetation washed up on distant shores. The ancestors of flightless birds that adorned the Pacific isles flew there first. Most of today's New Zealand wildlife did not survive the journey from Gondwana but arrived more recently and then evolved in subsequent isolation.[10] The flow of species around our planet is far quicker than the movement of continents, albeit still orders of magnitude slower than today's human-assisted torrent.

New Pangea is an apt metaphor for the accelerated connections of the modern world. By moving species across the surface of the planet, we are bringing about biological collisions every bit as significant – and just as permanent in the history of life on Earth – as when continents have in the past collided, or species have occasionally floated across the oceans. We are reuniting the biological world. Quite how far along this road we have already travelled is open to debate, but 971 bird species have been released in at least one location where they might be deemed to be 'alien' introductions between the years 1500 and 2000.[11] This amounts to approximately 9 per cent of all birds already, and the rate has been accelerating dramatically through time. A full quarter of all documented introductions were carried out in the last seventeen years of the five-hundred-year study. A project called DAISIE (Delivering Alien Invasive Species Inventories for Europe) has come to comparable conclusions. It established that Europe contains just over twelve thousand 'alien' species so far.[12] This amounts to nearly 10 per cent of the total current land and freshwater species in Europe, though the figure is much higher for plants (over a third of the species currently growing in the wild are international success stories that started their lives elsewhere) than for animals (about 4 per cent). Even this is bound to be an underestimate because the lists of imported species are nowhere near complete for fungi and for the smallest insects. Given that the flora represents the base of the ecological food chain, we can expect that the diversity of insects, fungi and microbes associated with each plant will catch up over the coming centuries.

This process is incredibly fast. Contact between Europe and the New World marks the emergence of genuinely global trade, the time when the transfer of species really started to increase. It has taken about five hundred years to get to where we are now, and each century has seen more movement than the one before. At the current rate of transfer, it might take another millennium or so to complete the job. This would be sufficient time for new arrivals to turn up and for recent colonists to spread more widely within the continents where they have become established. Assuming that the transport of goods and people continues, the geography of the world's animals and plants will have been thoroughly rearranged in less than two thousand years. Geologically speaking, this is virtually instantaneous. We are in the middle of the biggest biological pile-up in world history, an indelible signature of the Anthropocene.

We are reuniting Pangea in less than two thousand years, approximately 200 million years ahead of schedule.

Despite my scepticism about their attitudes to the arrival of new species, ecologists and conservationists do have entirely valid concerns. When species become established for the first time in parts of the world where they did not originate, there are bound to be repercussions. There are good reasons why *some* foreign species have gained a bad reputation. A combination of invasive humans, dogs and Pacific rats set off in boats from Southeast Asia around 3,500 years ago and spread across the Pacific in three waves of invasion. This deadly trio of big, medium-sized and small predators presented the natives of the islands they invaded with insurmountable challenges, generating ecological havoc in far-flung islands from Hawaii in the north to Easter Island in the east, eventually settling in New Zealand just seven hundred years ago.[13] The local birds, snails, insects and plants had not met intelligent, ground-living mammalian predators since their ancestors had colonized the islands millions of years previously. A mass extinction ensued, extinguishing virtually one in ten of the world's bird species[14] – a topic I return to in the next chapter. This was followed in recent centuries by the arrival of Europeans with guns, accompanied by a crew of two even more tenacious species of rat, cats, stoats, mosquitos, bird malaria, snakes, ants and predacious snails, among a menagerie of new invaders, which were fully capable of polishing off yet more of the native islanders.

Such losses are not entirely confined to remote islands. Worst of the continental invaders has been the 'chytrid' fungus *Batrachochytrium dendrobatidis*, which is now one of the world's most widespread and successful species, thanks to pregnancy tests. Before the invention of handy pregnancy-test strips, finding out whether you were pregnant involved a trip to the doctor, urine sample in hand, so to speak. The doctor then asked the advice of *Xenopus*, the South African clawed frog, which in its time had represented a revolution in pregnancy testing.[15] The samples were sent off to the labs, where a technician would inject the woman's urine into the back leg of a female frog. If it contained sufficient human chorionic gonadotropin hormone, the hapless frog would have ovulated by the morning – indicating that the woman was pregnant. Unfortunately, *Xenopus* turned out to be a carrier for the chytrid skin fungus and the pathogen was washed out in waste water into the world's ditches, streams and rivers; it may also have hopped off on the backs of a few escapee frogs. The best guess is that *Batrachochytrium dendrobatidis* originated in Africa but then romped its way through the skin of the world's frogs and toads, threatening hundreds of species. Some of the frogs and toads that it infected, such as the harlequin frogs of Central and South America, were highly susceptible,[16] and disappeared, whereas others had some resistance, and survived. Bad things can occasionally happen on continents, too.

The calamity of the Pacific, the loss of hundreds, if not thousands, of fish that once lived in land-locked African lakes, which have been disrupted by imported predatory fish, and the demise of chytrid-infected harlequin frogs in the New World, help explain why foreign predators and diseases have gained such a terrible reputation. Invasive species (if we include ourselves as hunters) have probably been the largest single cause of vertebrate extinctions since humans spilled out of Africa into an unsuspecting world. Given a choice, I would happily repatriate some of the most inconvenient of these super-successful species and resuscitate those that have become extinct. But this is not the world that we inhabit. Furthermore, invasive species are no longer near the top of the list of threats,[17] perhaps because the isolated ecosystems and species that were particularly susceptible to interlopers from distant lands have, in the main, already been invaded.

More remarkable than the losses, however, is quite how few foreign species have caused any others to go extinct at all. Most of the Pacific extinctions were accomplished by the aforementioned trio of big (human), medium-sized (dog) and small (rat) predators, and the rest by the arrival of a mere handful of additional carnivores and diseases, such as avian malaria and the mosquitos that transmit it. Only a few dozen out of thousands of foreign species that humans have transported across the constellation of Oceania's islands have driven any native species extinct. Even fewer have caused other species to become extinct on the world's largest continents.[18] With notable exceptions, including the chytrid diseases of amphibians and chestnut blight in North America,[19] foreign species hardly ever cause native species to become extinct from entire continents.

This is not surprising. Researchers who study the interactions between different organisms long ago demonstrated that a small number of species have a major effect on other species, but that most do not. This was first appreciated in 1969 by North American ecologist Robert Paine, who spent his days scrabbling around on rocky shores near the University of Washington, where he worked. He discovered that one species, which was a relatively rare starfish, had a huge effect on almost all the other species on the shore.[20] He christened it a 'keystone species'. When he experimentally removed this mussel-guzzling starfish, beds of mussels grew in great abundance, crowding out other rock-encrusting animals and plants that used to grow there. The entire biological community changed. However, although Paine is deservedly remembered for the keystone concept, it is worth recognizing that most of the other species he experimentally manipulated had very little effect – otherwise, the starfish would not have seemed remarkable.

The ever ebullient Dave Raffaelli, enthused by this research, set out to test Paine's ideas in the estuary muds of the Ythan River, north of Aberdeen in Scotland – a landscape so bleak and windswept that it perhaps accounts for his decision to grow a wind-cheating protection of dense facial hair. Disappointed not to find a keystone species in his initial work, he hit on a plan: to add and remove as many species as he could, in turn. He reckoned that persistence would pay; eventually, he would find the keystone species. After a decade or more of experimentally adding and subtracting species in search of the elusive keystone species, he gave up. There wasn't one. Lots of species had relatively small effects, a conclusion that is critical to our understanding of ecosystems. Sometimes there are one or two species in a particular place that have large and disproportionate impacts on all the others, like Paine's starfish, and sometimes there are not. Most species that are added to or subtracted from an ecosystem have little impact on the others.

This is equally true of foreign species, caricatured by York biologist Mark Williamson's 'tens rule',[21] whereby only about one in ten species that arrive in a new part of the world escape from captivity or gardens, only about one in ten of these then become fully established in the wild, and only about one in ten of the established species go on to be regarded as pests or weeds. A tenth of a tenth of a tenth. This ratio varies quite a bit from place to place, but roughly one in a thousand species that arrives causes a real issue for the native animals and plants, consistent with Paine's keystone ideas in community ecology and Raffaelli's failure to find any such species. And when people say that these species become pests or weeds, this usually just means that they become common, without actually endangering other species with extinction. Foreign species are acting like any other species: a few have major impacts, but most don't.[22] Because a large majority of them have such limited impacts, the importation of lots of new species almost always increases the numbers of species in any given location, just as we saw in the forests and waters of Lake Maggiore. When lots of new arrivals establish breeding populations, hardly any 'natives' die out as a consequence.

This is true even in New Zealand, one of the world's hotspots of extinction. When my wife, Helen, and I accompanied invasion biologist Jacqueline Beggs for a jet-lagged view over New Zealand's largest city, Auckland, we could almost have been at home. Species-rich meadows now cover the tip of the old volcanic cone of Mount Eden – Maungawhau, to give it its Maori name – that protrudes above the city, forming a vegetation comprised largely of European plants. Meadow plantains, wild cranesbills, sorrels and European grasses generated the scents of our own hay meadow. But walk into the remaining areas of forest near Auckland, and hardly a foreign plant species can be seen. A few introduced plants do live in the forest:

Kahili ginger from the Himalayas, check; wandering Willie, or *Tradescantia*, from South America, check; *Plectranthus* blue spur flowers from South Africa, check. But even though these plants grow in the forest, the forest is still dominated by native New Zealand trees, shrubs and ferns, and there is no indication that they will cause native species to disappear.[23]

This story is told and retold. On average, for every new species that arrives, less than one of the species that was originally there dies out.[24] The arrival of foreign plants has not only approximately doubled the diversity of New Zealand's flora,[25] it has also done so in the Hawaiian islands and elsewhere in the Pacific. The botanical diversity of these islands seems to just go up and up and up as new species arrive, with no obvious limit in sight.[26] This is also true of vertebrates. Despite the extinction of many native birds from the Hawaiian archipelago, the islands are now home to many different introduced mammals, lizards, frogs and freshwater-fish species, as well as to imported birds.[27]

This all leads to an increase in the diversity of each region. Take Britain, which now has six species of wild deer, rather than the original two. Some 1,875 foreign species of plants and animals have established wild populations in Britain in the last two thousand years, and mostly in the last two hundred; yet, as far as we know, no native species has died out as a consequence (although some have died out for other reasons).[28] This pattern repeats itself across the world's continents.[29] American states have experienced approximately 20 per cent increases in plant diversity through imports, and a similar level of increase has taken place for fish in American river catchments.

It is worth reflecting on the British ratio of 1,875 arrivals to zero extinctions caused by invasive species. With odds that low, I might cease to worry so much about legislating against new arrivals. Of course, some of the new arrivals do engender ecological changes – the replacement of one introduced crayfish by another, for example – that fall short of the extinction of native species, but change is how the biological world works.[30]

Many of these successful invaders have simply filled ecological voids and seized new opportunities that have been created by humans. We have changed the world's habitats: creating forests where there were grasslands, meadows where there was once forest, and cities along the coast. The burgeoning human-mediated traffic of microbes, fungi, plants and animals is accelerating the rates at which these human-made habitats are filled with species. The New Zealand forest is still largely populated by plants that have lived there for millions of years, whereas Auckland's human-created meadows and suburbs are dominated by species from Europe, Australia and Asia. Woodlands in England and in New England are often composed of native trees, shrubs and under-storey herbs, whereas human-derived habitats are dominated by plants that originated from further afield, be they from sand dunes along our coastlines, mountain crags or imports from other continents. In Missoula in the American state of Montana, the university campus teems with imported European birds, while the forest behind is inhabited by natives. Other opportunities arise because we have altered the Earth's climate, and incomers that we have transported from warmer regions are able to thrive under the new conditions. The biological world is on the move, taking advantage of new human-created opportunities.

However, ecological and evolutionary voids also exist for other reasons. For example, many oceanic islands were too remote to be colonized by land mammals, and their arrival has filled a void. And the dearth of broad-leaved evergreen trees in Europe was due to the history of the ice ages; introduced species have now plugged this gap. The myriad species that are now filling these historical voids and taking advantage of the new opportunities we have created are the initial inheritors of a human-altered planet.

Maggiore is emblematic of this new world. The Swiss Riviera may not be typical of the whole world – no place is – but there are many common themes. The consequences of hunting, agriculture, forestry, industry, gardening, fishing and tourism, not to mention climate change, are all evident in this one landscape. The numbers of species in the landscape have increased with the new diversity of habitats: the Italian sparrow, whose existence is a consequence of human arrival, is nesting in the eaves of the lakeside villas; yellowhammer buntings, clovers and comma butterflies are found in the farmland; green woodpeckers, European pond tortoises and cranesbills inhabit the golf course; yellow whirls of Cornelian cherry flowers, chirping tree sparrows and sulphur-powdered brimstone butterflies adorn hedgerows, while hawfinches, blackish-red squirrels and lime trees are found in the forest. The numbers of species have also increased with the importation of thousands of garden plants, and the lake is full of fish that have only just arrived. In addition, the warming of the climate has allowed the broad-leaved evergreens to make their home in the surrounding forests. The world has become a melting pot – New Pangea – transforming our planet in a few short centuries. And one of the consequences is that European forests of the future will contain more species, not fewer.

That's not so terrible. We have to be realistic and accept the world's biological systems for what they are. In the long run, it is the species that keep moving and successfully exploit new environments that will survive and prosper and thus ensure the survival of their kin on planet Earth. Successful species will continue to inherit the human-altered Earth.

PART III

Genesis Six

Prelude

In Part II, I concentrated on ecological change. The exploitation of animals and plants by humans, conversion of habitats, climate change and the transport of species have increased the abundances and geographic distributions of many species, as well as triggering the decline and extinction of others. They have generated declines in diversity in some locations but increases in many others. These ecological changes are at the forefront of environmental thinking, and rightly so. However, humans have initiated a period of evolutionary change that is just as fundamental as the ecological transformation of the world – in the long run, more so. When the environment changes, life ultimately responds by evolving. Evolution is how life on Earth comes back from disasters.

The 'Big Five' mass extinctions to befall Earthlings – those occasions in the last half billion years when three-quarters or more of all the species that previously existed became extinct – led to periods of biological recovery and diversification over the ensuing millions of years. Each mass extinction gave rise to its own genesis of new life-forms that would go on to dominate successive spans of our planet's history. Amid a new human-created mass extinction – some say mass extinction number six – we should consider whether we are on the brink of a sixth major genesis of new life. This is the topic of Part III.

The initial phase of recovery from the Big Five mass extinctions involved some creatures becoming successful while others dwindled. We might just think of this as ecological change, but the characteristics of species that enable them to be successful under new conditions are the products of past evolution. Birds and mammals were successful, whereas the much larger dinosaurs died out at the end of the

Cretaceous period. This represents evolutionary change on the grandest of scales. So, in Chapter 6, I contemplate the characteristics of some of the winners and losers of the human epoch. This evolutionary replacement of some kinds of species by others is already an indelible signature of the Anthropocene epoch.

Those species that do survive have begun to live under new physical and biological conditions. Every population of every species has experienced changes, be that to the climate, the acidity of the ocean, the levels of carbon dioxide in the atmosphere, increased levels of nitrogen in the soil and water, the habitats they live in, or the arrival of new invading species from across the seas. Given that genetic variation exists for nearly everything, evolutionary change inevitably follows, as populations and species respond to these new challenges. These are the events that I contemplate in Chapters 7 and 8. In Chapter 7, I evaluate how some individuals and populations are surviving better than others when conditions change in places where they already live. In Chapter 8, I examine how evolution is accelerating in the Pangean archipelago as species arrive in a new part of the world and meet other species for the first time.

Chapters 7 and 8 also reveal something more surprising. New species seem to be coming into existence with immodest haste, adapting to new conditions. Increasing numbers of species that are global adventurers are coming into contact with distant relatives, as we will see in Chapter 9, and hybridizing with them. These events are generating novel evolutionary forms and, on occasion, new species. The Italian sparrow is one such. Remarkable as it might seem, new plant species may be coming into existence faster today than at any time in the history of our planet. A new era has arrived in which we see an acceleration of evolutionary change and the genesis of new lifeforms. Given that many of them would not exist but for humans, they challenge us to contemplate the relationship between humanity and nature.

6

Heirs to the world

When the famous naturalist, writer and zoo-keeper Gerald Durrell arrived in New Zealand in 1962 in search of the elusive takahe, he and his film crew headed for the one place where a small surviving population hung on, in a valley high in the Murchison Mountains in the remote south-west of the South Island. Drenched in the pouring rain, and frozen to the core, Durrell temporarily forgot his discomfort, so astonished was he to encounter a multicoloured, scarlet-beaked, goose-sized bird popping out from behind a clump of snow grass. And then Takahe Valley, where they were, once more disappeared, as Durrell later recalled, under the *'muffling grey paw'* of encircling rain clouds, at which point he and his crew retreated to their mountain hut *'sipping whisky and tea in equal proportions'* to restore their inner warmth.[1] He continued: *'In the many years I have been hunting for animals in various parts of the world, I can never remember being so acutely uncomfortable as I was during our sojourn in Takahe Valley.'* And that is what saved them. The birds, that is – alas, it was the whisky that accounted for Durrell in the end. The environment was so harsh that it was even worse for the rats, cats, stoats and other introduced predators that kill takahes than for the flightless birds themselves. The endangered, plodding birds had just hung on.

The Murchison Mountains was the last place they survived, the final remnant population belonging to one of the two species of takahe that were entirely confined to New Zealand. Predators and then competition from introduced deer brought their numbers down to little more than a hundred individuals by the beginning of the 1980s. They were on the way out, until an intensive conservation programme was launched. Deer were shot from helicopters to reduce competition for grazing, a captive breeding project using takahe-shaped glove-puppet feeders with fake scarlet beaks (to stop the chicks imprinting on humans) was developed to bolster numbers, and then hand-reared birds were released on to safe offshore islands, where the predators had been trapped and poisoned. Since then, they have been established on seven such islands. Only around two hundred and eighty individuals exist, even now, but at least their numbers are going up and it is possible to see them again.

Fifty years after Durrell's visit, my wife and I were back in New Zealand. Annabelle, our volunteer guide, beckoned us past the toilet block, behind the café and visitor centre on Tiritiri Matangi Island, a climatically benign sanctuary in the Hauraki Gulf near Auckland. And there they were, half a metre high, gleaming blue-black and greenish feathers, beaks like the ends of a bright red anvil: a pair of takahe. It was a creature I thought I'd never see. But if the somewhat bow-legged, lumbering parents were astonishing enough, the chick was something to behold: fluffy, grey-black and reminiscent of a miniature B-movie tyrannosaurus in its unsteady pursuit of its parents. No match for a rat or a stoat. Turning back the clock to a mammal-free New Zealand seems to be working, in the sense that these heavyweight relatives of moorhens, coots and swamphens can survive and breed again in the lowlands, provided that they are maintained in places where predators are kept away.

One might reasonably ask how come they were on the lawn near the toilet block rather than in the 'native bush' that everyone was trying to restore? The answer, it seems, was that the gangling mini-monster and its corpulent parents were fattening themselves on a diet of the nutritious European-origin grassland plants that dominated the lawns, eschewing the tougher native vegetation. The flora that the takahes have been tucking into has more in common with my own meadow 18,000 kilometres away than it does with the remnant native bush a mere ten metres from where the birds were feasting. Not all foreign species are bad for takahes, and the removal of predators is only a façade of re-creating the past. From the takahe's perspective, successful alien mammals are terrible but successful alien plants are delicious. Another odd feature of this remarkable project is that there is no historical record of a takahe having ever lived on Tiritiri

Matangi, and those that live there today are South Island takahe rather than the now-extinct and larger North Island takahe, which used to live on the neighbouring mainland. Both species of takahe did live in the lowlands previously, but not, as far as we know, on Tiritiri Matangi itself.

Some will hate this, saying that it is like a zoo, with predators controlled, foreign plants to eat and the 'wrong species' of takahe introduced to a new island home. But, in a world where everything is somewhat changed, it is reality. We have no option but to get used to it.

The challenge is that some of the world's species are rising to the top, whereas others are losing out. The takahe is one of the losers, while rats, stoats and their ilk are thriving. This is a difficult problem. The history of life on Earth is one long story of successful animals and plants replacing those that proved to be less successful. Should we just let these winners replace unsuccessful species, or intervene so that the losers might live a little longer?

Nowhere is this quandary greater than in New Zealand, where the human inhabitants have decided to take action on behalf of the losers. They are aiming to slow the rate at which continental species arrive through customs, to wage war on the 'foreign species' that have already become established in New Zealand and to nurture every precious individual of their rarest and most endangered species. This is understandable. New Zealand is full of unusual island forms that do not live anywhere else, and many of them are a source of national and cultural pride. They are also of considerable scientific interest. Because New Zealand is a large area of land that has remained above water for a long period of time (it is a fragment of continental land), it represents the last stand of many ancient groups, including parrots (a group that contains the alpine kea, forest kaka and the nocturnal flightless kakapo) and tuatara reptiles, which used to be more widespread in the world. It is the only place where they have not yet been replaced, in the same way that lemurs survive only in Madagascar. Faced with so many endangered and unusual species, the human inhabitants of this island nation today dream of removing the mammalian predators their ancestors imported so that the 'rightful' inhabitants of the land can be saved. And they have started.

The idea is to eliminate rats, mice, feral dogs, possums, cats, stoats, weasels and the odd Mrs Tiggiwinkle so as to re-create a mammal-free world on large numbers of New Zealand's offshore islands, of which Tiritiri Matangi is one. In so doing, New Zealand conservationists have become world leaders in extermination. Setting off in helicopters and planes, they have air-dropped all-purpose mammal poison across entire islands, killing off any pests that take the bait. The aerial bombardment is followed by on-the-ground maintenance of poison stations and traps to ensure that the islands become, and then stay, completely predator free.

Not content with killing mammals on offshore islands, New Zealanders have started to build predator-proof fences around some areas on the mainland, and then they kill off the rodents, mustelids (ferrets, stoats and weasels) and possums that were unfortunate enough to be living inside the cages. The mammal-killing public have even assembled themselves into volunteer groups to poison and trap uncaged areas. As these areas lack fences, this requires continuous effort – every weekend spent killing mammals in the bush – to maintain a cordon sanitaire around locations where endangered New Zealand animals and plants might again thrive. The aspiration is that New Zealand's unreasonably large snails, giant insects, ancient tuatara reptiles and native plants will all benefit from this war on mammals, but it is the native birds that have really captured the public imagination. As soon as an island, mainland cage or uncaged death zone is declared predator free (or the number of predators sufficiently reduced), the local citizenry want to reintroduce as many of New Zealand's endangered birds as possible.

Much as I admire the endeavour, and appreciated the opportunity to see the galumphing birds, the whole approach suffers from three major drawbacks. The first is that many of the birds that used to live in New Zealand before humans arrived are extinct already, so it will never be possible to restore the original biological communities. The second is that it will not be feasible to rid Tiritiri Matangi Island, or anywhere else in New Zealand, of every insect, plant or fungus that originated in one of the world's other continents. In terms of the whole fauna and flora, the removal of terrestrial mammals only scratches the surface. The third, and most problematic, challenge is that all the effort merely staves off failure. A few of the world's most successful animals can be kept at bay for a while, but not for ever. We can kill boat-hopping rats and swimming stoats the moment they dare to place their paws on predator-free islands, but if humans were ever to depart New Zealand or get fed up with trying to save takahes, those islands would be reinvaded within a few decades or centuries. They are close enough to the mainland for stoats to swim back across to them, and all the 'protected species' would die out again. The challenge of saving takahe, kakapo, tuatara, mouse-sized crickets and saucer-like snails has been postponed, not solved. This is the fundamental difficulty for all conservation programmes when some species (the threat) are more successful than others (the threatened). They are saved only for as long as active intervention continues.

Conservationists recognize that reinvasion by predators is a problem, and they also appreciate that each offshore island (and onshore cage) is too small to support a viable takahe population. They are committed not only to guarding against reinvasion but also to moving animals back and forth between islands for ever more, so as to turn many tiny enclaves into one reasonable-sized population. Both of these problems would go away, however, if there were no mammals on the mainland either. There would be no rats left to reinvade the offshore islands and, better still, it would be possible to release walking birds and dopey reptiles on the mainland once more, where they would be free to wander off and reclaim their native lands. And so it has come to be that the new national strategy is to make the whole of New Zealand 'predator free'. This cunning plan has a slight flaw. No one knows how to eliminate the offending mammals from the entirety of New Zealand, or how to keep New Zealand predator free once they have.[2] It might be more realistic to release takahes on more modestly sized and remote islands elsewhere in the Pacific, in areas which have lost their own flightless birds and where the risk of reinvasion by predators is lower, or on some of the colder islands of the southern oceans, where rats have already been exterminated.

The most impressive of these projects, so far, was inspired by the success of New Zealand conservationists, but it was not carried out in New Zealand. Between 2011 and 2015 Tony Martin, a rugged animal conservationist from the University of Dundee's Centre for

Remote Environments in Scotland, spearheaded a project that involved massive air-drops of poison to remove rats and mice from the 3,700-square-kilometre island of South Georgia.[3] This appears to have been spectacularly successful, although follow-up surveys are still required to be certain that no pockets of population have survived. Not only is this the largest area cleared to date, but South Georgia's minimal human population and remote position in the southern Atlantic Ocean make biological defence of the island far more feasible. Yet, even in this brief five-year period, one subsequent ship-borne rat incursion has taken place and had to be wiped out. This reminds us that we have taken on a for-ever fight. Whatever the scale of a particular project, the problem remains. Any project that re-establishes an unstable ecological system will have to be defended indefinitely against reinvasion by continental beasts.

In the unlikely event that every rat and mouse can be removed from New Zealand, the challenge of stopping rodents from ever reappearing will be humongous, unless all transport of goods and people to New Zealand is outlawed. New Zealand was first invaded by the kiore, or Pacific rat, which was whisked from its Southeast Asian homeland to islands throughout the Pacific Ocean, until the kiore was replaced by the smart black rat, which purloined food stores from our houses and barns in tropical Asia before it was transported across the planet. The black rat was in turn replaced by the scabby-tailed brown rat from northern China. Having unleashed the rodent hordes, people have subsequently done their best to stem the tide, to little avail. History tells us that there is really only one effective way to keep any particular species of rat at bay over a large geographic area – and that is to wait for an even more pugnacious species of rat to arrive and wipe out the previous pest. Kiore, black rat, brown rat: which one next? There are over six hundred mouse- and rat-like rodent species in the world (as well as many more other rodents). There is a long waiting list, and we have no particular reason to suppose that the global succession of rodent life has finished just yet.

Rather than attempt to assuage our ancestral guilt and defend an unending siege, it might be better to go with the flow. If we step away from the presumption that the old species are better than the new and must be saved, it is evident that – if we add up the mammals, birds, reptiles and amphibians – there are more different kinds of verte-brates living in New Zealand today than there were before humans arrived, and there are twice as many plants as before, as well as hosts of imported insects. Most of them are not going to go away, even if a few of the introduced species can be extirpated completely and others removed from specific locations. If we want to have oceanic islands that contain flightless birds, either out of a sense of nostalgia or because we think that they will help maintain the vegetation (which evolved with flightless birds for many millions of years), then they need to be animals that can survive in the presence of rats, cats and other successful continental species. It does not make much sense to try to restore ecological communities with species that pass away at the merest whiff of a stoat.

We should look to the continents where there are plenty of walking birds that still have the wherewithal to deal with mammalian preda-tors. We could introduce cassowaries, rheas and emus to New Zealand to compensate for the missing giant moas.[4] We could also encourage the growth of populations of turkeys, chickens, quail, pheasants and other game birds, which are already living in the wild there. These are feral birds that walk while they are foraging but evolved on continents that are full of predatory mammals, so they retained the capacity to take off when attacked. Bustards, tinamous and flight-capable geese are also realistic options. Ideally, those that are released would be species that are themselves threatened, like Brazil's endangered Alagoas curassow, protecting the introduced spe-cies at the same time as restoring walking birds to New Zealand ecosystems.

Another strategy would be to introduce diseases, predators or com-petitors that would reduce the densities of the introduced mammals to the point that they would no longer kill off the native species. Strictly vegetarian rodents, for example, might be able to displace the more omnivorous rats and mice from most rural and forested habi-tats. New Zealanders are understandably nervous about this, given that some of the predators they dislike so much were originally intro-duced in previous failed episodes of biological control. Otherwise, attempts could be made to modify the native species so that they will be able to survive in the presence of introduced mammals, potentially using new genetic technologies, engineering them so that they would breed more rapidly, nest in safer places, defend their nests aggressively, run faster or be more resistant to new diseases. For example, rather than continuing the existing policy of killing the naturally occurring hybrids between Australian pied stilts and New Zealand black stilts, it might be preferable to use these hybrids to breed a new race with the physical appearance of the black stilt but with the ability of their Australian cousin to avoid being killed by predatory mammals. In other words, turn losers into winners.

Irrespective of the merits of any of these particular suggestions (which should proceed only following carefully controlled trials), there is a broader logic. Rather than always try to defend the losers, we could seek to build new biological communities composed of compatible species so that future ecosystems are more robust than those that currently exist. In the end, it is ineffective to put all our efforts into schemes that work against the natural grain of the biological world. Backing losers may be honourable in intention, but backing winners will be more effective.[5]

The most spectacular success story of the last million years has been the evolution of a fast-running, group-hunting, intelligent, tool-wielding ape – nothing like a human had existed before and, consequently, we have changed the world. Apart from species that have been directly killed by humans or eliminated because we removed their habitats, the remaining losers mainly come from places where the biological world was in some senses 'incomplete', in that particular types of animals or plants which are found throughout most of the rest of the world were 'missing' until humans came along and introduced them, and the introduced species then filled the void. These biological exchanges reveal 'what works best'. When it comes to vertebrates, it is clear that small and medium-sized land-dwelling mammals are particularly effective. The moment they arrive in places where they were previously absent, they initiate biological revolutions that are difficult or impossible to reverse.

Despite their poor reputation, rodents are intelligent, resourceful animals; cute even, with their large eyes and ears, twitching noses and elegant whiskers. Rodents represent evolutionary designs that have been extremely successful in the past – with 2,250 species in existence,[6] they represent half of all mammal species – and their ability to survive and diversify shows no signs of waning. Humans have been able to extinguish most land mammals that weigh more than a tonne, but we have made only modest inroads into those that are lighter than a kilogram. This is good news for rodents, and good news for mammals in general: 70 per cent of all mammal species are rodents, bats and relatives of shrews, hardly any of which weigh more than a kilo. Giant hundred-kilogram flat-tailed beavers and capybaras are among the few rodents that did disappear when humans spread out across the world, but nearly all the smaller ones survived. There is no prospect of humans exterminating every species of rodent.

Equally, sparrows and their relatives are doing disproportionately well – we could never extinguish all the 5,400 species of perching birds that exist. Individual species may vary in their fortunes, but the overall design of passerine birds is not under threat. They are evolutionary success stories, with or without humans. The same is true of lizards, frogs, fish, beetles, butterflies, trees, grasses and lichens. The ordinary evolved designs of animals, plants, fungi, bacteria, archaea and viruses that have been successful for many millions of years, and that are the most usual forms of life throughout the world's continents, have coped perfectly well with the arrival of humans, just as the survivors of previous mass extinctions were ordinary species that would have been considered completely unremarkable before those calamities. They were just minding their own business – eating, reproducing, fighting off diseases and trying to avoid being eaten – before the world was turned upside down. The heirs to the world already surround us. However, when we move these really successful types of species to locations where they did not previously exist, there are casualties.

Given that most small mammals and birds seem to be coming through the Anthropocene relatively unscathed, it is perhaps easier to identify the characteristics of the losers; the winners are the remainder, by default. Despite their overall success, some small mammals have disappeared already. The big-eared hopping mouse, which used to bounce its way across sand dunes in Western Australia, was last seen in 1842, close to Perth's Moore River. This may well be sheer bad luck. Any species, whatever its biological characteristics, is prone to extinction if it is confined to a small part of the globe and humans transform all of that area of land in a manner they are unable to adjust to. This is why it is so important to protect examples of all the different kinds of habitats that exist in the world, especially in places where there are concentrations of species that live nowhere else.

However, it seems not to be just about bad luck. It is small mammals that live on islands, rather than in small parts of continents, which are consistently in the greatest trouble. Christmas Island's corpulent bulldog rat, for example, shuffled off this mortal coil in the early years of the twentieth century (it was last spotted in 1903), perhaps unable to cope with the arrival of more agile ship rats or the diseases they incubated. We have to face the fact that some extinctions take place because the species that arrive are simply 'better' at carrying out particular ecological roles than the previous residents, or they are superior in their abilities to kill and avoid being killed. This is how evolutionary replacement works. By moving species from one continent to another, and from continents to islands, humans have accelerated the process by which the eventual winners come out on top.

New Zealand bats, having arrived in the land of the long, white cloud, where there were no ground-dwelling mammals to compete with or attack them, undertook a change in lifestyle. They started hunting for insects and other invertebrates on the forest floor. They evolved a shuffling-swimming motion to crawl through the leaf litter; an ingenious way of folding their wings to stop them being an impediment on the ground; extra talons for walking and catching prey on the ground; and a special arrangement of muscles and tendons to help them take off (much like the extinct flying pterodactyls).[7] While these are all fascinating adaptations, the basic problem is that, in becoming evolutionary masters of the sky some 50 to 60 million years ago,[8] bats long since 'compromised' their ability to walk, sacrificing their fingers to support expanses of skin. The result of this evolutionary digit-conversion is that New Zealand's greater short-tailed bats walked on the wrists of their forearms, and these are just not as good as hands and feet for pedestrian activities. The rats have it.

Furthermore, heavily pregnant flying mammals are not airworthy, so bats produce only one offspring each time they breed. For the greater short-tailed bat, this amounted to one pup a year, in contrast to the brown rat, which can become sexually mature when it is about five weeks old, has a gestation of three weeks, and litters that contain seven and sometimes even more kittens. This allows the rat population to at least triple in two months when food supplies are plentiful, whereas it would take the bats three years or more to achieve the same level of population growth. Thus rats have a much greater capacity to take advantage of temporary gluts and to bounce back from times when their numbers have been reduced by predation. Rats, stoats and cats were a fatal combination for New Zealand's greater short-tailed bat, which was last seen in 1967. RIP. Conservation efforts are now under way to save the lesser short-tailed bat, which typically spends rather less of its time hunting on the ground and is therefore somewhat less susceptible to ground-dwelling predators. It would certainly be a shame to lose such an evolutionary oddity.

Sad as this may be, it is hardly surprising. And the future evolution of the world's mammals is not threatened by the loss of ground-dwelling bats. The walking bat was an eccentricity that was unlikely to catch on, even if the vampire bat can do it, too, hopping around the legs of animals whose blood it wishes to drink. In contrast, bats undeniably outperform rats in the sky, and their ability to echolocate – to bounce high-pitched sounds off moths and other nocturnal insects and listen for the echo – makes them better than most birds at hunting in the dark. The global supremacy of bats as airborne night-time hunters is not under threat.

At least the bats still had some capacity to fly, whereas many of the island birds faced an even greater evolutionary challenge. The ancestors of all birds sacrificed their front legs over 70 million years ago,[9] when their forelimbs evolved into wings. Wings have great advantages. They enable birds to obtain food that cannot be reached from the ground, to find new places to forage and to escape from ground-dwelling and tree-climbing predators by taking to the air. Wings make sense. However, birds use up enormous amounts of energy to grow, maintain and use their flight muscles, so flight can be a disadvantage when there are no predators to escape.

If you live on a remote island where the only predators are crabs that can be evaded by walking away, any ground-feeding bird (as opposed to those that feed in trees or catch insects on the wing) that puts its energies into flight is likely to leave fewer offspring than its waddling relatives. So, time after time, flying birds colonized isolated oceanic islands and then started to lose their power of flight in worlds of mammal-free bliss.[10] They evolved into stocky beasts, with big legs for walking rather than wing muscles to take off. This is what happened to the takahe. Its ancestors were flying swamphens of the genus *Porphyrio* which settled in New Zealand about two and a half million years ago and proceeded to evolve into the heavy, flightless birds that we see today.[11] Island birds also lost their fear – seals and sea-lions were not going to chase them. The story is similar with disease. Why expend metabolic energy preparing to fight off bird malaria and other non-existent pathogens?[12] Any birds that redirected this energetic expenditure towards other bodily functions would be at an advantage. In the end, the world's archipelagos were filled with oversized, tame, disease-prone, slow-breeding, flightless birds.

Such was the situation when humans turned up in New Zealand about seven hundred years ago, accompanied by our personal menagerie of continental carnivores. Rats and pigs would have depleted the birds' food and plundered their nests, while humans and our canine friends could easily polish off the adults. It was far more probable that heavyweight, tame, walking birds would become extinct than that they would remaster the air. By sacrificing flight, they had ended up with only two functional limbs, and they were no match for agile four-limbed mammals. Takahes were not capable of defending themselves, even though their ancestors, other *Porphyrio* species that lived in Australia and elsewhere, did have this capability. Rats, dogs, pigs, stoats and humans won.

Flight-worthy island birds have also declined in more recent times, mainly after the arrival of European sailors and settlers who brought cats, mice, new species of rats and sundry other animals to torment the natives. New Zealand's grey-and-rusty-red kaka parrot nests in tree holes that are accessible from the ground, making them vulnerable to clambering rodents, Eurasian stoats and Australian possums. Chestnut-winged New Zealand saddlebacks roost at night on branches near the trunks of trees, which was a perfectly sensible

The pukeko (above) colonized New Zealand only five hundred to a thousand years ago. It can still fly, and continues to thrive, despite the presence of predatory mammals. In contrast, its relative the takahe (below; seen here on the island of Tiritiri Matangi) arrived about 2.5 million years ago and evolved into a heavy and flightless bird that is incapable of surviving predation by ground-living mammals.

strategy until nocturnal tree-climbing carnivores arrived – why would they have sat on inconveniently small twigs if nothing was coming to get them? Then we inadvertently released a variety of pathogens and mosquitos that initiated epidemics on many islands. Introduced continental birds acted as carriers, which prevented the native birds recovering once the first outbreak had taken place – the demise of two-thirds of the flight-capable Hawaiian honeycreepers can be credited to their inability to fight off avian malaria, in addition to their susceptibility to predation by mammals. A capacity to fly was not sufficient protection. When we look across all the heavyweight, predator-naive and disease-prone birds that used to exist, there is no single cause of death. It was the combination of the characteristics of the island birds that did for them.[13]

While it is a great pity that a thousand species of island birds have disappeared, this event may not have much, if any, effect on the long-term evolutionary future of birds. Nearly all these island forms were ultimately descended from flight-capable, predator-savvy and disease-resistant birds that started life on the world's continents.[14] Once stranded, they became suited to mammal- and disease-free worlds, making them unable to survive on continents ever again, or to survive on islands once continental species arrived. Eventually, the island-adapted birds would have disappeared, either as the islands ducked back under the waves, or as fresh continental species arrived and displaced them.[15] By mixing up the world's species, humans have accelerated their demise rather than altered their eventual fate.

The flip side is the increased success of those species that originated on the world's continents. Flitting white-eyes, for example, still peek out of shrubberies in many of the Pacific and Indian Ocean islands. Having colonized the world's islands 'only' in the last million or so years, most of the island white-eyes could still outwit feline killers and deal with continental lurgies, when they arrived. The New Zealand pukeko has also survived perfectly well. Just like the original takahe, it is a *Porphyrio* swamphen, but it arrived from Australia only about five hundred to a thousand years ago, and so it can still deal with 'outsiders'. Pukekos can fly, usually have four to six eggs in a clutch instead of the takahe's two, and they continue to thrive in the presence of predatory mammals. The takahe represents New

Zealand's past, and the pukeko its future. Loquacious myna birds, red-whiskered bulbuls and a host of other birds that originated in the world's continents now greet visitors to numerous oceanic islands. As a consequence of these introductions, the diversity of birds on most of the world's remote islands has increased over the course of the last century or two. Birds, as a whole, are doing well, even if a substantial minority of the species that used to exist in the pre-human era have disappeared.

The consistent failure of island forms provides insight into what types of animals are most successful. It is no great scientific revelation. Successful species need to be resistant to disease and have the ability to avoid predators, and they must have the capacity to reproduce fast enough to replace individuals that die – abilities which apply to nearly all continental species. There are no great surprises, either, when we contemplate the mechanical design of land-dwelling vertebrate animals: four-legged mammals are effective at living on the ground and clambering up trees, birds move through the daytime air and bats hunt in the night-time air. Of course, this is a simplification (owls hunt at night, for example, and ostriches walk), but this is a true reflection of the design features that have worked best for the last 50 million years. It is no different now. In the absence of four-legged land mammals and the insect vectors of disease, the rules used to be different on oceanic islands. Now they are not. Continental life has been spreading throughout our island realms, with the consequence that already successful animals and plants just became a little bit more so.

Now that we have investigated the state of island diversity, the next challenge is to understand what happens when species move from one continent to another. There is no better place to contemplate this than in a tropical forest, the ultimate representation of successful life on Earth. I first visited the tropics in 1979, although a day later than originally planned, having accidentally flown to Panama City in Florida en route to my intended destination of Panama City in the Republic of Panama. Not put off, I was soon sweating my way up the endless flight of steps that led away from the small dock on Barro Colorado Island. Before I knew it, a chattering white-faced capuchin monkey had appeared. Then a brown-throated three-toed sloth was spotted, hanging like a slow-moving sack in a tree. Now surrounded by water, Barro Colorado Island is a former hilltop that became marooned when Lake Gatun rose around it in 1913, the lake having been created as part of the Panama Canal that connects the Atlantic and Pacific Oceans. Barro Colorado Island sits bang in the middle of the Isthmus of Panama, which became a crossroads of biological life when the enormous, continent-sized island of South America 'bumped into' North America several million years ago. The connection of the Americas permitted bears, cats, gomphothere elephants, peccaries, tapirs and the ancestors of llamas to walk southwards into South America for the first time, and monkeys and sloths to clamber through the trees in the opposite direction. Armadillos, opossums and porcupines also journeyed northwards towards their Central and North American dream – as did ground sloths and armoured glyptodonts, which were later exterminated by humans. Pausing for breath on the steps of Barro Colorado Island, I stood on a piece of land that had allowed tapirs, peccaries, capuchin monkeys and arboreal sloths to move through just a few million years ago, and they are still there, living together.

The exact sequence of events remains somewhat disputed, mainly because there was no single day on which the Isthmus of Panama miraculously emerged from the waves. The gap between north and south became an archipelago of emerging islands as the Cocos geological plate in the Pacific ground its way beneath the Caribbean plate (then mainly land with decreasing gaps), until a continuous strip of solid ground was achieved perhaps 3 million years ago.[16] This lack of a single contact date makes interpretation of the fossil record rather tricky. It has been further complicated by the continuing trickle of animals from Asia into North America over the last 2.6 million years, when sea levels dropped sufficiently during successive ice ages for Siberia and Alaska to be periodically joined together into the Beringian Plain. It has been a time of increasing connection between the biological worlds of South and North America, just as we are now seeing human-mediated increases in the connections between continents, and between continents and islands.

Whatever the exact timing of each crossing, far fewer of the southern mammals made it into North America than vice versa, with modern cats, including the present-day jaguar and now-extinct sabre-toothed *Smilodon*, replacing the marsupial cat-like animals that used to live in the south. Perhaps the self-sharpening (by moulting the old surface as a sheath), fast-action retractable claws gave cats the edge over their marsupial equivalents. Perhaps it was their ability to produce more offspring at once, or their resistance to disease. Whatever it was, the large land animals that originated in North America, some of whose own ancestors had previously arisen in the connected continents of Africa, Europe and Asia, thrived, whereas those that originated in the somewhat more island-like continent of South America fared less well once the Americas were joined together. Giant ground sloths and some of the other South Americans were highly successful in North America, but, nonetheless, the species that originated in the large northern continents were the ones that were more likely to come out on top: at the time when humans first arrived in the Americas, half of the genera of mammals in the south were of northern origin, whereas only 20 per cent of those in the north had originated in the south.[17] In other words, the northern mammals were two and a half times more successful as colonists. It seems as though the animals that evolved in places where the fight for survival had previously been keenest – in the larger continents, where there were plenty of open habitats where these types of species thrived – won the evolutionary contest. South America was a bit more island-like than North America.

When humans arrived in the Americas about fifteen thousand years ago, it was again the large animals that had originally evolved in South America that were most susceptible. If we look at today's paltry remnants of the original fauna, it is no evolutionary surprise that South America's largest survivors – the tapir and llamas – evolved in predator-rich North America, and that the jaguar, the largest surviving mammalian predator, had ancestors that are thought to have originated in the Old World. The 'evolutionarily savvy' species whose distant ancestors emerged in Asia and North America were simply more effective at coping with the challenge of sophisticated predation by armed, ground-hunting humans. The best-equipped species survived.

In stark contrast, the vegetation of South America and tree-living animals were not overwhelmed by invaders from the north. The flow of rainforest species tended to be in the opposite direction. The tree sloths and capuchin monkeys that welcomed me to Barro Colorado Island evolved in the south and spread northwards towards Mexico (which is geologically, and was biologically, part of the north) when the continents collided. This is consistent with the same idea. Most of the tropical forests of the Americas were in the south, and Amazonia remains the largest area of tropical forest in the world to this day. For these species, South America could be thought of as larger than North America, and these were the species that were most successful in the warmer and wetter parts of the northern continent. The consequence was that, despite the success of ground-dwelling mammals from the

This male jaguar specializes in killing caiman and capybara that haul themselves up on sandbanks along rivers in Brazil's Pantanal. Jaguars are the largest surviving cats in South America. Marsupial lions disappeared from South America at around the time that modern cats and the now-extinct sabre-toothed cat, Smilodon, *arrived.*

north, the overall rate of migration from South to North America was approximately 30 per cent higher than the rate of movement in the opposite direction.[18]

*

A similar pattern applies to species that move between continents and to those that move from continents to islands. Species that originate in vast expanses of a certain type of environment seem slightly better equipped – on average – to 'win' the ecological and evolutionary contest when they come into contact with species from smaller areas. This makes Australia particularly interesting. The continent of Australia and New Guinea (which is one landmass during ice ages, when sea levels are lower) is not physically connected to the other continents and can be thought of as somewhere in between an island and a continent. Furthermore, Australia is biologically unique. Most of Australia's ground-living and tree-climbing mammals were marsupials until humans arrived – not forgetting the odd egg-laying platypus and echidna. Giant grazers, top carnivores and medium-sized omnivores, right down to the smallest mammals, were marsupials. But, unlike in New Zealand, rats arrived about a million years ago, and mouse-like rodents before then; and there were bats, of course. Small placentals had already arrived.

When humans colonized Australia approximately fifty thousand years ago, we hunted two-and-a-half-metre-high giant kangaroos, two-tonne wombats and the rest of the mega-marsupials to extinction – just as we exterminated mega-placentals elsewhere. Marsupial lions disappeared around the same time. However, we did not bring new species of mammal with us on that occasion. Dingos were not imported until some forty thousand years later, and predatory foxes, cats and ferrets were not released until Europeans arrived, when we also brought over dromedary camels, water buffalo, horses, sheep, cows, rabbits and sundry other placental grazing and browsing mammals. The stripy, dog-like thylacine vanished during the European era, while the stocky, fierce devil is extinct from the Australian mainland but still survives in Tasmania. However, this is not necessarily because they are marsupials: ten of the sixty-four species of native Australian placental rodents became extinct, too, many Australian marsupials still survive, and opossums live in South and North America, where the mammal fauna is dominated by placentals.

Whatever the specific combination of causes, successful mammals of European and Asian origin – the largest landmass – have replaced a number of their Australian counterparts. On the other hand,

Australia's ground-living birds were able to cope quite well. One-and-a-half-metre-high blue-necked cassowaries and grimy-brown emus survived their encounter with Eurasian mammals.[19] Australia has long been full of animals that are fast, venomous and dangerous, which has resulted in emus and cassowaries evolving not only the ability to sprint away from predators but also the capacity to deliver unpleasant blows and even to eviscerate unwary mammals that attack them. These defences proved sufficient for them to cope with the arrival of human hunters and the other imported carnivores, in stark contrast to New Zealand's apparently quite similar-looking moas, which disappeared almost as soon as humans colonized the islands.[20]

The rest of the Australian birds continue to fly. If you go camping in the Cape York Peninsula of northern Queensland, keep your eye out for brush turkeys that come looking for scraps. Brush turkeys have vulture-like necks of vermilion-red skin, a vivid yellow collar and a black body that ends with a tail that has the weird appearance of having been ironed into a vertical position. Approach them, and they scuttle off into the undergrowth. But, if pushed, they explode into the air in a somewhat ungainly manner, flying up to land on a branch that is out of reach of the potential predator. They and many other Australian birds seek their food on the ground, but they have retained their ability to escape from ground-living predators when required.

Whether Australian mammals could invade the world's larger continents is largely untested. Brushtail possums are thriving in New Zealand despite the authorities' attempts to exterminate them, and there seems to be no particular reason why they would not be successful elsewhere. Small populations of several kinds of wallaby are living in New Zealand, on Oahu in the Hawaiian Islands, in the British Isles, and in various other locations around the world. There is even a wild-living wallaby population to the south-west of Paris, in France. However, few introductions have been made of Australian mammals to continental areas where they really might be able to thrive. The larger and faster kangaroos are perhaps the most interesting because the kangaroo design has no equivalent among large mammals in the rest of the world. Kangaroos lived with marsupial cats for millions of years, out-hopped humans for fifty thousand years, and then managed to flee from dingoes for thousands of years. They might well be able to outwit predators on all the world's larger continents, given the chance. Their mode of locomotion, combined with an impressive ability to survive and reproduce in regions with sporadic rainfall, is yet to be tested in the grasslands and savannas of South and North America, in the Mediterranean region or on the Indian subcontinent. I put my money on the kangaroos.

While species that originate from larger, more diverse and better connected locations tend to win the evolutionary struggle, there are also plenty of successful species that originate in 'smaller lands', although they are, in most cases, species that come from large and biologically diverse[21] habitats within these more island-like places. Species from the vast rainforests of South America were successful further north, and it might be presumed that Australia's dryland plants and animals, including kangaroos, could be effective elsewhere – Australian wattle trees are already spreading across parts of southern Africa. New Zealand is renowned for its extensive cool rainforests, which support some two hundred species of earthworm that live nowhere else in the world. It is perhaps not surprising, then, that a flatworm predator of New Zealand's earthworms is now happily consuming European species, following its introduction to the cool and damp British Isles. New Zealand is, perhaps, like a mini-continent for the enemies of earthworms.

It should also be remembered that most species are not steam-rollered out of existence when biological worlds collide. The majority of all species in every region remain quite successful, presumably because their prior experience of competition, predation and disease, and their adaptations to local conditions, enable them to ward off most incomers. New Zealand still has nearly all its original plants, most of its invertebrates and about half its birds. Australia is still mainly populated by species with a long history of living there – and the highly successful passerine birds seemingly originated there before colonizing and diversifying in the rest of the world. Likewise, South America is still full of species whose ancestors evolved in South America millions of years before it became connected to North America. Subsequent evolutionary diversification of rodents within

Australasia and South America has more than made up – if we simply count up the number of species – any deficit associated with the initial extinctions.

Great replacements have frequently been at the heart of large-scale and long-term evolutionary change, most of which took place so long in the past that we now think of the consequences as simply the way the world is. Standing in the humid tropical luxuriance of Barro Colorado Island, I was in a place that bore witness to such events. I was at the crossroads between the southern and northern continents, where one of these great transitions unfolded. It was also the place where humans passed through as they moved into and once more transformed the biology of South America. In more recent times, the Panama Canal has increased the flow of marine life between the Atlantic and Pacific for the first time in millions of years. Wherever one looks around the world, humans are accelerating the transport of species from one location to another.

Mix the species up and see who wins. Based on the lesson of the coming-together of South and North America, and of Eurasian animals arriving in Australia and New Zealand, it will be species from the most biologically rich parts of the world (for each particular type of environment) that are most likely to thrive. But we can also see, both today and in the geological past, that the overall consequence of these biological exchanges is to increase the diversity of life within each region.[22]

In the middle of a period of change, as we are today, it is perhaps natural that the emphasis is often on the losers. If we were to look back on today from many millions of years in the future, we would see that humans caused an increase in extinction. The Anthropocene would represent an epoch that saw the final extinction of ancient groups which had hung on in remote locations where they had not yet been displaced. But we would also see the success of plants and animals that will form the basis of life in those distant post-Anthropocene times.

A geologist 10 million years hence will notice how the world's geography changed in a remarkable manner during the Anthropocene epoch. Individual species that were previously restricted to single continents became global in their range and, by this point, they will have diversified in their new homelands. A future geologist would remark on some great replacements, and on others that received the final nails in their coffins. Unless we prevent it deliberately, the Anthropocene will be the last stand of the tuatara, a primitive group of reptiles that arose about 220 million years ago but proved incapable of surviving in the presence of a full set of modern vertebrates.

Yet the geologist would also see that, apart from several groups of exceedingly large animals, which humans hunted to extinction, nearly all the major types of animals and plants that were widespread before the Anthropocene epoch would have survived. The most successful will have been completely ordinary, everyday species: ground-dwelling mammals, flying birds, bats whizzing past in the night air. The heirs to the world are not bizarre, weird things and ancient evolutionary relics – those are the species that are dying out. The success stories are already all around us. Look out of your window and the chances are you will be staring at the future.

7

Evolution never gives up

The Generals Highway snakes its way through a forested land interspersed with boggy meadows and rocky domes in California's Sierra Nevada range. Black bears root their way through delicate, pink shooting-stars in the meadows, and lightning-struck pines stand as sentinels on high ridges. The highway ushers travellers between the Grant and Sherman groves of giant sequoias, where the visitors stop and gaze upwards at the reddened, fissured trunks of the most massive plants in the world. They enjoy the splendour of grand landscapes and experience a wilder, unchanging world. But this is no pristine landscape. Loggers moved through the forest in the late 1960s, leaving behind clearings of powdery granitic soil where the trees failed to regenerate. It was on one of these dusty slopes that I found myself sitting in 1984 – contemplating why it was absolutely teeming with Californian butterflies, while the same species was far rarer in the less disturbed parts of the forest. We do not normally expect species to be commoner in places that have been so seriously 'damaged' by human actions.

As I sat there, Edith's checkerspot butterflies fluttered by on patterned wings, a mosaic of rusty-orange, black and creamy-white flecks. Every so often, a female landed inexpertly and laid a batch of eggs on the delicate blue-eyed Mary plants; when the eggs hatched, the caterpillars would consume the foliage. Despite their enthusiasm for depositing eggs on blue-eyed Mary plants that were growing in the dust, elsewhere the checkerspots laid their eggs on louseworts, which are distinctly less attractive-looking plants with grubby-yellow flowers that grow in the forest's relatively undisturbed dappled shade. In choosing blue-eyed Mary plants, the butterflies in the human-altered habitat seemed to be behaving differently from those in the surrounding landscape.

To find out if this was true, my PhD supervisor Mike Singer and I decided to ask the butterflies themselves. How this worked was that Mike wandered around wielding his butterfly net, and from time to time brought me female checkerspot butterflies that he had caught either on the dusty slope or in the butterfly's original habitat. Meanwhile, I sat cross-legged, cramped, back aching, notebook open, a pyramid-shaped cage placed on the ground in front of me. To either side were spherical butterfly-containing cages, known as Singer balls, so named because of my supervisor's own bodily deficiency and because Mike had designed them. I took each female butterfly out of one Singer ball, and then placed her on blue-eyed Mary plants in the pyramidal cage. If she liked what she tasted (through chemical receptors on her front legs), she would curl her body into a semicircle and settle down to lay a batch of eggs. Then I quickly removed her before she had had a chance to deposit the first egg, so that she would still be ready to lay if offered another plant. Once all the females had been tested, I gathered up my cages, stretched my legs and walked over to the nearest lousewort. There I sat down again and repeated the whole exercise, this time placing each female butterfly on the lousewort. After that, I went back to the blue-eyed Marys and did it all again. Then back to the lousewort, and so on.

Because each butterfly was numbered with a Sharpie pen, I could track exactly how each female had behaved over a number of days. Lots of them would curl their body around the lousewort leaves very readily, and they would do so every time I placed them on this plant. But if I placed the same butterflies on a blue-eyed Mary in between times, they would just sit there and sun themselves. This would go on for hours, and sometimes for days. Clearly, they liked the lousewort better. This was not particularly surprising because lousewort was the plant that had traditionally supported the development of checkerspot caterpillars along the General's Highway. However, some of the females seemed to like the blue-eyed Marys just as much, and a few of them liked them better than the lousewort. The individual female butterflies had their own personal opinions of how much they liked to lay their eggs on the two plants.

Edith's checkerspot. Female number 6 is laying a batch of yellow eggs at the base of a lousewort (below). Another female is laying on an introduced white man's footprints plant, Plantago lanceolata *(above). Eggs can be found on this plant (opposite, top), and the caterpillars eat it (opposite, middle). The adult female (opposite, bottom) shows the endangered subspecies, Taylor's checkerspot, which survives mainly because it has switched its diet to feed on the introduced plants.*

Mike Singer and his students were witness to an amazing event. The butterflies were in the process of evolving a love of blue-eyed Marys in front of our eyes.[1] Those that Mike had caught in the undisturbed dappled shade mainly liked to lay their eggs on louseworts, especially where the plants grew below scattered pine trees that surrounded natural rocky openings in the forest. But those from the disturbed dusty slope were perfectly happy to lay their eggs on blue-eyed Marys. It transpired that their caterpillars also survived better when they were eating this plant, so females that were prepared to lay their eggs on it, and carried genes to do so, left more offspring – they had evolved a new egg-laying behaviour that increased the consumption of the blue-eyed Mary plants.[2] In less than twenty years, foresters had cut the trees down, butterflies had moved in, the butterflies had evolved to use the plants that were now growing in the new forest clearings and, by the middle of the 1980s, the checkerspot population had built up to record numbers. They had become adapted to a human-created habitat.

Something even more impressive was taking place on the other side of the mountains, conveniently close to Carson City, to which we could retreat whenever it snowed or we felt the need to avail ourselves of the cheap but excellent buffets that were provided to entice potential gamblers into the casinos. What better reason could there be to visit Carson City? Naturally, we were there to study butterflies that were in the process of evolving to live alongside humans.

A century earlier, cattle ranching was the prime activity in Nevada, before the state built an entertainment and tourism business out of activities that are commonly known as 'vice' in the rest of the country. While loggers converted the forests in the Californian sierras into new, open habitats, ranchers were transforming the meadows of Nevada. A host of European hayfield plants sprouted in the wake of the settlers. Among these was ribwort plantain (alternatively, long-leaf plantain), known as 'white-man's footprints' to the indigenous human population who had previously hunted and burned the land. White-man's footprints is an appropriate name. These plants initially followed the clearance of forest in Europe and became exceedingly common in meadows across large swathes of the continent. Centuries later, the whitish people from western Europe inadvertently took the plantains with them as they colonized the world, most likely trans-porting the seeds in hay that had been brought to feed their livestock. White-man's footprints thus continued its global journey, arriving on the volcanic cones of Auckland in New Zealand as well as in the meadows of Nevada – a botanical sparrow. Plantain plants popped up all over the place, and especially where the meadows were irrigated to generate fresh, lush growth to nourish herds of cattle.

From the perspective of the checkerspot butterfly, there was an advantage to any female that laid her eggs on this newly arrived plant. The plantain leaves remained green long enough for the caterpillars to survive during dry summers, which seemed to be getting a little drier with the first signs of climate change. In contrast, the native plants they used to eat shrivelled up and most of the caterpillars starved or desiccated. With this difference in survival, the butterflies started to evolve a liking for laying their eggs on plantains: the proportion of female butterflies content to lay their eggs on this plant rose from under a third in 1984 to three-quarters in 1987.[3] A few years later, the switch was complete.[4] Human-caused evolution was in full flow.

Years later, I found myself contemplating this experience as I strolled around my thinking meadow in Yorkshire. Back in the 1980s, lying in a tent in the Sierras listening to the howl of coyotes echoing across the mountain slopes, I presumed that such rapid evolutionary events were unusual. But I was beginning to wonder if they were. If rapid evolution is rare, what are the chances that two out of a few tens of populations that Singer and his students studied would be evolving so rapidly? Was this incredible good luck? Or are bursts of rapid evolution far commoner than we realize?[5]

As I pondered the fate of checkerspot butterflies, I also reflected on the voluminous updates Mike had emailed me from time to time over the ensuing years. Neither of these particular evolutionary episodes worked out. Freak spring weather and a decline in blue-eyed Marys caused the butterfly population to crash along the General's Highway, and the remaining butterflies went back to feeding on louseworts. This was not an evolutionary failure because the population still sur-vives, but neither can it be counted as a new success. The Carson City episode came to naught as well – the population died out when

Freak weather in the Sierras, in California. A younger version of the author in the middle of the checkerspot season, with adult butterflies waiting to be tested for their choice of plant. The butterflies are being kept in 'Singer ball' cages, which are hanging in the trees.

management of the meadow changed. We had witnessed rapid evolution, but neither event generated long-term change.

But evolution never gives up. It only needs a very small proportion of these spurts of evolutionary change to work out and become new biological success stories. When we look across the entire range of Edith's checkerspot, we can see that it has switched its diet to feed on ribwort plantain elsewhere: in Oregon and Washington states, and in neighbouring British Columbia in Canada. The federally endangered Taylor's checkerspot (a subspecies of Edith's checkerspot, whose historical habitats have been lost) is so reliant on it that conservationists are actively planting plantains out into the wild.[6] To provide a supply of butterflies, prisoners at the Mission Creek Corrections Center for Women in Washington state breed checkerspots in a greenhouse so that they can be released into these new habitats. Odd as it might seem, actively encouraging an alien plant (increasing gains) is helping to conserve a much-loved native insect (reducing losses).

Meanwhile, back east, many populations of the related Baltimore checkerspot eat plantains, too.[7] Across all the populations throughout the geographic ranges of all of the checkerspot species in North America, it is virtually guaranteed that some of them will eventually flourish on this introduced plant and start to spread – even if many other populations die out because they fail to adapt fast enough to changing conditions. Given that white-man's footprints now grows right across the world, these plantain-eating checkerspots have enormous potential to prosper in the future.

It is not just checkerspots that are experimenting. The caterpillars of one third of all of the 236 native butterfly species that live in California include newly arrived exotic plants in their diet.[8] That is astonishing, given that most of these introduced plants have been growing in California for less than two hundred years. American butterflies are seemingly rushing to exploit foreign plants. In Europe, where fundamental habitat change has a longer history, nearly all butterflies make use of human-modified land. In the picturesque Dordogne valley in France, five species of fritillary, all distant relatives of North American checkerspots, can be found gliding on ochre wings across traditional meadows that, without humans, would be forest. Not only are they living in a human-created habitat but, like their North American cousins, the caterpillars of three of the five consume the same kind of plantains.[9] The plants would not be there without people to maintain the meadows. Perhaps these European checkerspots evolved to make use of them thousands of years ago, when the forests were first cleared.

It is the same in England. Continuing my stroll, I notice a brown argus butterfly. This insect was concentrated in the south of England in the 1970s and has spread northwards only in recent years. The males perch in sheltered corners, glistening in the sun, hinting at an ultraviolet sheen I cannot see. They lustfully intercept passing females and fly out pugnaciously if a rival intrudes. They, too, have had a change of diet. Working with my brother, Jeremy, butterfly conservationist Nigel Bourn established that the caterpillars of this insect eat juicy rockrose plants, which mainly grow on hot slopes – on south-facing hillsides – on the chalk-and-limestone hills of southern England.[10] Hot caterpillars grow fast, allowing the butterfly to complete its lifecycle – from egg to caterpillar to chrysalis to adult butterfly – twice each year. Then, as the climate began to warm, it became possible for them to complete their lifecycle on flat terrain. The only catch was that rockroses were few and far between, away from the chalk hills. They were stuck. And then the brown argus did something very similar to the checkerspots. In fact, it was even more impressive because it allowed the butterflies to double their British range in less than twenty years.

Rather than place their eggs on the usual rockroses, some of the female brown argus butterflies took to laying on small wild geranium plants. In particular, they liked one called *Geranium molle*, known by the inconveniently cumbersome name of dove's-foot cranesbill. Laying eggs on this plant proved to be an enormous advantage because, unlike the rockroses, these particular geraniums are common in the British landscape. From the 1980s onwards, the butterfly expanded away from its traditional rockrose-containing suntraps on to road verges, field margins, country parks, rough pastures and even building sites, spreading northwards across landscapes that would previously have been regarded as too cold and lacking in suitable habitats.[11] In the early 2000s, they appeared in my meadow, 150 or more kilometres from where they used to live, not only laying eggs on wild geraniums that grow where the ponies churn up the ground but actually shunning the rockroses I have planted. The geranium-liking brown argus butterflies that live over large parts of northern England today are genetically different from those that used to live in chalk downland habitats in the south.[12] Just as some species are thriving under the new conditions of the Anthropocene epoch, some genetically distinct populations and individual genes are becoming increasingly successful. European butterflies did not evolve to make use of human-altered habitats thousands of years ago and then 'bask in their achievement'. They are still changing.

It is not just butterflies. Rapid evolution must be taking place in nearly all animals, plants, fungi and microbes. Almost all populations and species contain genetic variation, and some variants will survive better than others under the novel physical and biological conditions that now exist. The atmospheric concentration of carbon dioxide is already higher than it has been for some 3 million years, and may well end up higher than for 20 million years. This alters the capacity of plants to carry out photosynthesis, acidifies the oceans and warms the climate. Greenhouse gases have already raised the world's average temperature by 1°C and, writing in 2016, we are on course to reach an average global temperature that will be the highest for 3 million years,[13] and possibly for 10 million. Needless to say, temperature directly alters the physiology and growth of almost all organisms, hence their survival and reproduction. Humans have also changed the amounts of nitrogen, sulphur and atmospheric dust in circulation. And as we have seen, we have removed most of the largest vertebrates, transformed the land for agriculture, set species on the move as a result of climate change and generated a modern version of Pangea.

Faced with these novel conditions, it follows that nearly all wild animals, plants and microbes must be evolving in response. Genetically distinct individuals and populations differ, for example, in how fast they grow at different temperatures, their size, and in their interactions with other species (including the choices female butterflies make when selecting plants that will be good for their offspring to eat). There is variation in almost everything that anyone can imagine. When the environment changes, some of these variants are almost bound to survive at least slightly better than others, such that the characteristics of the next generation will differ from those that went before. Given how much and how fast humans have changed the world, it is entirely credible that we are now living through the most rapid period of evolution since the aftermath of the extinction of the dinosaurs 66 million years ago.

Of course, evolution is not a get-out clause that will enable every population or species to survive – populations of flightless birds do not contain any individuals that can soar off into the sky and escape from invading hordes of carnivorous mammals. Failed evolution will be commonplace. However, the populations and species that do survive will be somewhat genetically different from those that went before. This is the normal stuff of Darwin and Wallace's theory of evolution by natural selection. The difference now is that humans are spurring evolution on. When Charles Darwin first distinguished between evolution by natural selection (in nature) and by artificial selection (guided by human intent), he could not possibly have imagined that humans would change the world so much that we would influence the evolution of nearly every population in the world. It is worth considering artificial selection, then, to gain insight into the speed at which evolution can take place when humans are at the helm.

With one simple extension of his tongue, Rex could lick the entire surface of the kitchen table. Many was the time I would return to a smear across the table and a chewed cheese wrapper on the floor. Rex was our Irish wolfhound, a sighthound bred for strength and speed, thought to have been developed originally for hunting wolves and larger animals, and possibly for war. He was 85 centimetres tall on all fours, 2 metres on his back legs, weighed about 55 kilograms and ate a lot of cheese. He was usually extremely friendly, but his immensely deep growl and enormous teeth were sufficient to turn our soot-covered chimney sweep distinctly pale. While Rex was the only Irish wolfhound in the village, there were several Yorkshire terriers. At 3 kilos apiece and 22 centimetres high, 'Yorkies' are better suited to tackling mice and rats than wolves. Chihuahuas are even smaller, a mere kilo in weight and 18 centimetres high, perfectly adapted to living in celebrity handbags.

No one doubts for a moment that they are all dogs, or that they are derived from domesticated wolves. On the other hand, one wonders what palaeontologists would make of their fossils in the distant future. It is not so easy to decide whether any two types of (closely related) animal belong to the same or different species because the process of speciation is a continuous one of increasing evolutionary separation. We expect there to be situations where it is hard to call, one way or the other. However, a greatly reduced ability of one animal to reproduce with another is integral to the concept of a species. For this reason, fossil hunters have assigned the 4.3-metre-high giant mammoths on the North American mainland and the 1.7-metre 'micro-versions' from the Californian Channel Islands to different species. This is quite reasonable. The island mammoths were not only geographically separated but likely morphologically incapable of mating with their continental relatives, which were 2.5 times taller and about 12 times heavier. If we apply the same reasoning, a 4.7 times taller and 50 times heavier male wolfhound would require advanced yoga skills to reach down to a female Chihuahua, and even then probably not manage to get the appropriate part of his anatomy in the required location; if impregnated, half-wolfhound pups developing inside a Chihuahua bitch would probably prove fatal. Similarly, a male Chihuahua would need to levitate to reach a female wolfhound, unless she lay down. This is not completely impossible, but unlikely. Chihuahuas and wolfhounds are as different as many species of deer are different from one another, or lions and tigers, or cattle and bison. In many respects, they are even more different, despite the fact that dogs started to separate from wolves only some fifteen thousand years ago.

It is quite reasonable to argue, therefore, that Chihuahuas and Yorkshire terriers represent a different species from the original wolf. There are plenty of intermediate-sized breeds, however, and so it would be possible for the genes of Chihuahuas to make it into wolfhounds over the course of several generations and thence, perhaps, into wolves.[14] However, this does not normally happen, and the clubs that maintain pedigrees deliberately ensure that each breed remains as 'pure' as possible. Whether you take the view that some or all domestic dogs represent a species that is distinct from the original wolf is not the issue. The important point is that very substantial evolutionary changes can take place in short periods of time and can generate differences that are as large as those between species.

The pace of change is still quite slow, relative to the lifetimes and memories of individual humans, so we don't usually notice. Yet paintings and photographs of dogs from a hundred or two hundred years ago are often surprising. The bulldog used not to be nearly as squat, the pug not so snub-nosed and the Alsatian used not to have such a crouching gait. Once thick-set, snub-nosed and crouching were specified as the ideal standards of a particular breed, dog owners ensured that they selected individuals with these characteristics for their future breeding programmes. And so, generation after generation, the breeds have increasingly become caricatures of those original standards, success at dog shows having taken over from hunting ability as the agent of selection. Within a few hundred years – perhaps

1790

1890

2010

fifty dog generations – breeds have changed radically under the influence of selection. Over the full period of domestication, the largest have remained similar in size to wolves, yet Chihuahuas have become so small that their dimensions are comparable to those of fennec foxes, the smallest of all wild canids. The entire size range across all the wild species of wolves, dogs and foxes, which separated from one another over 20 million years ago, has been replicated in a few thousand generations of human-influenced evolution. Both the speed and the magnitude of change in modern dogs are remarkable. Evolution is not necessarily a slow process.

Dogs are but one example. Cattle, pigs and horses also show great variation in their size, while sheep and llama breeds differ in the character of their wool. Darwin particularly loved the varieties of the domesticated rock dove (the same species as the feral pigeon): appearance, behaviour and flight differ so much among them that the different forms could easily be taken for dozens of different wild-pigeon species. Similar changes can be seen in plants. One species of cabbage, *Brassica rapa*, has been bred into turnips (a root vegetable), Napa cabbage (Chinese cabbage), rapini (a broccoli-like vegetable), field mustard (an oil seed) and mizuna (a peppery salad). Different varieties of *Capsicum annuum* produce bell or sweet peppers that range from green and yellow to red and purple; other varieties are

Development of an increasingly squat body shape in the bulldog between 1790 and the present day. Modern breeders try to obtain pups from animals that best meet the breed specification. The UK Kennel Club 2010 'Breed Standard' specifies that the bulldog should possess the following attributes: 'Skull relatively large in circumference. Muzzle short, broad, turned upwards and deep. Flews (skin on the cheeks) thick, broad and deep, covering lower jaws at sides. Ears small and thin. Teeth not visible. Teeth large and strong. Neck thick, deep and strong. Shoulders broad, sloping and deep, very powerful and muscular. Forelegs very stout and strong, well developed, set wide apart, thick, muscular and straight, bones of legs large and straight. Chest wide, prominent and deep. Back short, strong, broad at shoulders. Hind legs large and muscular, slightly longer in proportion than forelegs.' *The breeders have been rather successful, although it escapes me why anyone would want a dog that shape.*

chilli peppers and jalapenos that vary in the concentration of heat-producing capsaicin chemicals (some associated with medicinal as well as culinary uses); while small, roundish-red 'Bolivian rainbow' and the dark-leaved and black-fruited 'Black pearl' are ornamental garden varieties that are not intended to be eaten at all. Without prior knowledge of their histories and genetic relationships, a botanist could easily assign these varieties to many different species, with the assumption that they evolved millions of years ago.

Darwin called all this artificial selection, but the processes are just the same as in any other form of evolution. Impressing a human admirer, and thereby gaining the opportunity to reproduce, is not fundamentally different from a peacock with an especially large and brightly eyed tail impressing an admiring peahen and thereby obtaining the matings that will pass on his tail genes. Nor is it fundamentally different from the butterfly that has passed on more copies of its genes because it laid its eggs on a different plant. There is no clear dividing line. The development of different varieties of animals and plants is simply a consequence of some individuals surviving and reproducing better than others, and thereby becoming highly successful in the Anthropocene. They succeed because they have characteristics that result in their genes being propagated. The fact that humans have played such a major role in this propagation is hardly surprising, given the worldwide abundance of humans and our impact on the Earth. Widespread animals will always affect the evolution of many other species, and humans have simply taken this to a new level.

Many of these relationships are mutually beneficial, just as they are between bees and flowers, and between fruits and birds. By visiting flowers, bees receive rewards of sugary nectar and nutritious pollen; while the flowers benefit from the bees acting as mobile reproductive organs, transferring some of their pollen from one plant to the next. Of course, there are costs. The flowers have to produce nectar and lose most of their pollen, and the bees expend energy transporting the pollen. Nonetheless, it is mutually advantageous because the genes of both plants and bees are passed on to subsequent generations at increased frequency as a consequence of the partnership. Similarly, birds benefit from eating berries, while the seeds of those plants are deposited in places where they may grow (often encased in a bag of bird-dropping fertilizer). Given enough time, and strong enough selection, extremely complex interrelationships can develop.

Mistletoes, for example, need their seeds to end up on the branches of the trees they will subsequently parasitize, which is why mistletoe berries have evolved to be sticky. A bird will swallow the fruit, digest the nutritious outer parts of the berry (benefitting the bird), then regurgitate it. The stickiness of the fruit requires the bird to wipe the seed off its bill, and where better to do this than on a small branch or twig, to which the seed sticks and then grows into a new plant (benefitting the mistletoe). Some Bornean mistletoes and flowerpecker birds have gone even further, opting for a rear-exit strategy.[15] These mistletoe seeds have tadpole-like tails and pass completely through the gut of the birds. The seeds then emerge at the bird's backside and get stuck, but once the birds wipe their bottoms on a convenient branch, the extremely sticky tail glues the seed in position. The gardener bird has planted the seed. The bird took advantage of the plant, and the plant took advantage of the bird.

This is no different to the relationship between us and our crops, or our livestock. We plant; they grow. Any genes our livestock or crops possess that cause humans to propagate them more effectively will increase in frequency; any variants of our genes that enable humans to make better use of these plants and animals will also increase within the human population. Take the ability to digest milk. Mammals do not need lactase enzymes after they have been weaned of their mother's milk, so they stop producing them. It requires energy to produce lactase, and thus it would be a disadvantage for adults to keep secreting it. This is how all humans used to be. Once our ancestors started to keep cattle and had access to milk, however, the benefits of being able to digest milk outweighed the metabolic costs of producing lactase. The adaptation to keep the genes for lactase 'switched on' in adults has subsequently spread through a third or more of the world's human population. Cattle have evolved in the presence of humans to produce more milk and meat, which has been a great evolutionary success for them (the worldwide cattle population numbers 1.5 billion); and humans have evolved an increased capacity to digest the dairy products we get in return. Successful cow genes to produce more milk and successful human genes to digest it have increased in tandem.

There are plenty of similar examples. People from populations with more starch (from crops) in their diets usually have more copies of a gene that produces salivary amylase to digest it; many other genes associated with the digestion of carbohydrates and fats have also changed since humans adopted agriculture. Even our tooth enamel and the way we sense flavour have altered as a consequence of mutualistic relationships between humans, our crops and our livestock.[16]

While these direct evolutionary relationships between humans and our domestic animals and plants apply only to a minority of all the species that live on Earth, they have reshaped our planet. Over a third of the world's land surface is covered by human-selected crops, pastures and forestry plantations,[17] and many of these plants bear little resemblance to their ancestors. Approximately 97 per cent of the total weight of all mammals added together is dominated by these newly evolved relationships. The genes of animals and plants that are favoured by humans have been successful, and the genes of humans that have enabled us to make a success of these partnerships have also grown in number.

Everything we are doing to the world is forcing evolution into overdrive.[18] The butterflies are evolving when their habitats are changed and when new plants arrive from the other side of the world. Our agricultural systems are dominated by human-altered animals and plants, as we have just seen. Whenever we kill individuals of wild species – whether for food or to eradicate pests and diseases – the survivors nearly always have characteristics that enable them to survive better than those which died. When we fish, evolution favours individuals that can escape through the mesh of fishing nets and breed before they are killed. Thus, individual fish have become smaller and breed faster (offsetting fishing deaths) in the great majority of populations where fishing takes place.[19] This evolutionary response may not be sufficient to prevent populations from declining, but at least it makes it somewhat harder for us to exterminate fish stocks entirely. Evolution can be good news, even if we would prefer our fish to be larger.

However, recovery of the survivors becomes a problem whenever we attempt to exterminate a species deliberately. 'Diseases of the past' are evolving resistance to antibiotics, while human hair and body lice, bedbugs, disease-transmitting mosquitos, sand flies and blackflies are increasingly difficult to kill with insecticides.[20] Our crop pests typically evolve resistance to new insecticides in a decade,[21] and over 200 different weeds have overcome at least 150 herbicides.[22] As for mammals, brown rats have repeatedly evolved their resistance to anticoagulant poisons, and mice are not far behind.[23] Each gene for resistance in all these pathogens, plants, insects and mammals is its own biological success story.

Evolutionary responses to climate change are probably even more widespread. In Finland, the pale grey plumage of owls that used to provide camouflage in snowy weather has been replaced by brown-coloured individuals that are hidden in darker, rainy forests;[24] and German blackcap warblers have abandoned their migration to the Mediterranean to take advantage of warmer British winters (and a nation of people who put out food for birds in the winter). Butterflies have grown large muscly thoraxes (facilitating flight) and bush crickets have developed long and powerful wings to help them speed towards the North Pole.[25] When it comes to biological invasions, the cane toads that are invading Australia have evolved longer legs that enable them to walk faster, accelerating their spread across the continent.[26]

Even pollution has been influential. Peppered moths evolved blackened forms that were camouflaged against the trunks of trees in polluted industrial Europe (they are now evolving to look like lichens again, since the introduction of clean air legislation),[27] while some species of plants have adapted to grow in soils that have been polluted by heavy metals.[28] And so it goes on. Innumerable evolutionary changes have taken place in response to climate change, to introduced species, to habitat change, to new poisonous chemicals, and to harvesting, and in most cases these changes have taken place on a timescale of decades. Human-influenced evolution 'in the wild' seems to be happening just about as fast as evolution by 'artificial selection'.

A truly global episode of rapid evolutionary transformation has been unleashed, as particular genes enable the individuals, populations and species that bear them to succeed. This should be no surprise. There are bound to be successful (and unsuccessful) genes, just as there are successful (and unsuccessful) species. Oddly, though, while the declines and increases of species are widely recognized, each case of evolutionary change is still reported with surprise. It is as if, ever since Darwin, we have swallowed the story that evolution is so slow that we can't usually see it. This suits science writers and natural historians, who can report each new example with a sense of shock and amazement; it suits scientists, who milk it for publicity; and it suits ecologists and conservationists, who find it convenient to treat species as if they are fixed entities. In the absence of national and global schemes that track evolutionary change – in contrast to the Audubon bird counts and other programmes that document changes to the abundances and distributions of species – each new instance can be revealed as a surprising event.

We all know perfectly well that evolution is not unusual, however. It is how life works. Genes are passed from every generation to the next in every population of every species in the world. Successful genes inevitably increase and inherit the human-dominated world, hence evolutionary responses to human-caused environmental change must already be virtually universal. Moreover, these changes are virtually instantaneous on a geological timescale. We live in a time of rapid and accelerating evolutionary change.

8

The Pangean archipelago

Sea lions lolloped on the beach while frigate birds soared overhead. The *Diamante* swayed on the gentle waves, anchored in the shallows while we came ashore to visit the island of Española. A family of primates – my wife, sisters, brother and sundry in-laws – relaxed under the tropical sun. I was sitting on a log that had washed up on a lonely beach a thousand kilometres away from the nearest continent, drinking water from a plastic bottle. Suddenly, a speckled-grey bird with an elongated tail reminiscent of a mockingbird hopped up beside me. It seemed to be eyeing my water bottle, and then it started to drink from my hand, its darting tongue tickling my palm. It was the tamest wild bird I had ever met, though not stupid – fresh water is in short supply on Española, a cactus-dotted patch of brown earth adrift in the Pacific Ocean.

Later that day, we left Española, the *Diamante* parting flocks of ocean-going red-necked phalaropes as we set sail into the evening light, white-vented storm-petrels dancing in our wake, seeking out the plankton that had been disoriented by the churning water. We woke the following morning to a furore of diving boobies, circling upwards, folding their wings and plummeting in shuddering beak-first dives into shoals of hapless fish, perhaps corralled under water by the sea lions, penguins and flightless cormorants that patrol the rich but chilly waters that flow between the islands of Isabela and Fernandina. Seas of plenty surround lava-strewn islands where life is harsh. Alighting on Isabela, we were again met by mockingbirds, but they were far less tame, darker around the face and back and had fewer speckles than those on Española. They are also genetically distinct. The Isabela mockingbirds are not the same species as those on Española.

Our journey continued in Darwin's footsteps, travelling from one Galapagos island to another, seeing new variations on a theme. Each island had mockingbirds that are obviously quite similar to one another, yet also a bit different. Some, it transpires, are slightly different forms of one species, some are entirely separate species, and others are hybrids whose ancestors had come from two or more other islands.[1] The differences exhibited by the mockingbirds across the islands impressed Darwin,[2] as did the differences between the giant tortoises: animals from different islands are distinguished by the shapes of their shells. On the drier islands, their carapaces form neck-accommodating arches, enabling them to reach up for food during times of drought; on the moister islands, their shell is conventionally tortoise-shaped, as green vegetation is always available on the ground.

Darwin was beginning to formulate the idea that animals would become adapted to new conditions whenever they colonized fresh locations. They would start to look different and, in the case of mockingbirds, behave differently from one another. This is not surprising. Soils in the region differ according to the type of lava from which each Galapagos island was formed and its geological age. Rainfall varies, too. The younger islands in the west have high volcanic peaks that force the moist air upwards towards the summits of the mountains, where the water condenses into clouds and rain falls. In contrast, the peaks have long since eroded on some of the more ancient, lower-lying islands, and they are prone to drought. And it is not just the climate. The small offshore rocky islands are more likely to be covered by nesting seabirds, their nutrient-rich guano peppering the ground. Then each island has a separate set of animals and plants, altering the food that is available and the shelter provided by the vegetation. These distinctions between the islands are sufficient to set evolution off on slightly different tracks: partly predictable because the characteristics of the islands differ, and partly governed by the chance genetic mutations that arise on each island.

Originally, a single species would have colonized one of the islands from a distant continent, Darwin reasoned, and then have spread from island to island.[3] The populations of this one species will then become somewhat isolated and experience varying conditions on

Rowing ashore from the Diamante, *we were first met by sea lions guarding the beach access to the island of Española* (top). *Once on the beach, extremely tame Darwin's mockingbirds hopped up and begged for water and scraps (juvenile* middle right, *with parent; and portrait of an adult,* bottom).

each island,[4] where they evolve different characteristics and eventually become two, three, or even as many separate 'daughter' species as there are islands. This is the point the Galapagos mockingbirds have reached, with separate species on some islands, and described varieties (subspecies) on others. One of the daughter species might, however, become so different from its ancestors – in the foods it eats, in the habitats it occupies, or in the way it recognizes its mates – that it could potentially now live alongside other daughter species. All it needs to do is colonize another island.

We know that the gap between the islands is not insurmountable – otherwise, the birds would never have flown to them in the first place – so the chances are that this will eventually happen. At some point, one of the daughter species starts colonizing the other islands, so that there are now two species living on each island, both descended from the same ancestor. This can potentially go on, again and again. The daughter species that has successfully colonized all the islands can now start to become a bit different on each island, just as before. And then they can separate into separate 'granddaughter' species, and the most distinctive ones may once again be able to spread back across the islands of the archipelago.

This is what happened in the case of the Darwin finches, the birds for which he became particularly famous; although, at the time of his visit, the young naturalist was more intrigued by the mockingbirds. Larger- and smaller-beaked species that predominantly eat larger and smaller seeds, respectively, can live together, having, it seems, originated separately on islands where the types of seeds that were available differed in size (depending on how dry the island is). Over the two – perhaps three – million years that finches have been living on the Galapagos, this process of separation and coming together, and of evolving to consume different foods and live in different habitats, has happened many times. The ground finches that greet you on Santa Cruz island search for seeds that have fallen on to bare earth, yet they live alongside small tree finches that dangle from bushes and the branches of trees, while the woodpecker finch can manipulate thorns and thin twigs to find insect food the other species cannot reach.[5] Naturally, the detailed ancestry of each species is complicated,[6] but the basic pattern is clear: diverge on different islands and come together. This is the way of archipelagos.

The Galapagos are not that unusual. Darwin would have been equally amazed if he had arrived in the Hawaiian islands. Stout-billed relatives of present-day Asian rosefinches colonized these islands about 6 or 7 million years ago and encountered an archipelago that was far wetter than the Galapagos.[7] The word 'colonized' makes this process sound purposeful but, more likely, a small flock was caught up in a violent storm, struggled on and finally collapsed, exhausted, on a distant beach. In any event, they arrived, found food and water, survived and started to breed. As they spread along the island chain, they diversified. Some became thick-set finches, others dainty honeycreepers. This diversification went on and on, until the original general-purpose finch had descendants with an enormous diversity of bill shapes and sizes suited to consuming nectar, fruits, insects and even snails. They became so different in their habitats, foods and behaviours that many species could live alongside one another on the same island. And the story is much the same for other Hawaiian animals and plants: the dramatic silversword plants and dancing fruit flies evolved unique forms on different islands and in different habitats, as did the happy-faced and web-building spiders, and Hawaiian geese, among many others. Another archipelago, another species generator.

Nowhere is separation followed by speciation easier than in archipelagos:[8] remarkably, 10 per cent of all land bird species were confined to Pacific islands prior to the arrival of land mammals, even though the total area of these islands represents just one quarter of 1 per cent of the Earth's land surface.[9] Archipelagos have been the perfect place for one species to turn into many, and for Darwin to develop his revolutionary thoughts.

This diversity-generating process is not limited to oceanic archipelagos. Lakes surrounded by land are islands for water-dwelling animals and plants – the wet archipelagos of lakes in the African Rift Valley are filled with a spectacular diversity of fish. Mountaintops surrounded by lower lands represent islands to animals and plants that are unable to survive in the hot lowlands. Similarly, base-rich calcareous soils may exist as isolated outcrops, and heavy-metal-containing serpentine rocks are likely to be surrounded by less toxic geologies. The story is similar. Diverge in different locations, and then there is the potential to come back together.

On longer timescales, this same process takes place at a planetary scale. After South and North America had gradually joined together and eventually become one continent, species that had evolved separately over very long periods of time came into contact. Although South American marsupial lions and some other species died out as a result of the contact, the total number of species increased.[10] First, this was simply because the number of immigrants into each continent was larger than the number of species that died out when the two biological worlds met. And then the number of species on Earth started to increase. Deer that had spread from North America evolved into fourteen species that are restricted to different habitats and geographic regions within South America (this is twice as many species of deer as exist in North and Central America put together).[11] Likewise, camel relatives that moved southwards evolved into vicuña and guanaco.

More dramatically, over 375 species in more than 80 genera of New World rats and mice[12] evolved after they negotiated the then archipelago of Central American islands and arrived in South America around 5 million years ago; and perhaps 80 species of pea-like lupin plants have come into existence within the last one and a half million years in different parts of the Andes, a mountain range that can be thought of as an archipelago of cool, open habitats at the highest elevations, above an ocean of lowland trees and deserts.[13] Going in the opposite direction, the Virginia opossum, North American porcupine and armadillo came into existence, their predecessors migrating from the south to the north.

As we saw in Chapter 6, continents are not entirely isolated, even when there is a gulf of water between them. Some movements are quite regular. Dust-like bacteria, fungal spores and the minute seeds of orchids can be blown huge distances, and tiny insects such as thrips and aphids can be transported as aerial plankton in the wind. Migrating birds today regularly cross the Caribbean Sea, with the odd seed, spore and louse on board, and they must have passed back and forth between North and South America long before the two continents merged. For most animals and plants, however, the transport of species between continents is exceptional. Nonetheless, just as freak storms may very occasionally deliver exhausted birds, seeds or reptiles clasping mats of floating vegetation or logs to remote islands, they have the potential to deliver individuals from one continent to another. Cattle egrets seemingly did manage to fly from Africa to South America in the 1870s. Although the movement of individuals may be rare, those that survive the ordeal will eventually evolve into new species in their new home. A few tens of millions of years later, their descendants might reverse the journey.

Distant parts of the same continent can also be connected, isolated and then connected again. Asia and Europe form a single landmass, Eurasia, but the oak trees and beech trees that grow in eastern Asia and in Europe are not the same as one another. A single species of beech tree forms cathedral-like forests carpeted in golden autumnal leaves in Europe, but different beech species grow in China and elsewhere in eastern Asia, some with trunks that soar towards the canopy, while others branch close to the ground and jostle for space with bamboo thickets on mountain slopes. The climate of central Asia is too harsh, and these trees survive best in the more moderate oceanic regions that exist towards each end of the Eurasian landmass. Isolated, they have become separate species. European oaks also differ from those in eastern Asia, as though the two regions were giant islands separated by the frigid aridity of the continent's centre. North American beeches and oaks differ again from those in Europe and Asia. Perhaps 55 million years ago, a new kind of tree evolved: the first oak. It originated, spread, colonized different continents, evolved into different species in different regions and climates, and at least some of them came back together again; acting like a giant, slow-acting global archipelago.

The rate at which species came back together and fuelled new episodes of diversification in the global archipelago operated so slowly, relative to the timescales of human generations and civilizations, that we could be forgiven for thinking of these global-scale patterns of biological diversity as unchanging. But this is no longer the case. Today, humans have become that long-distance glue that has turned isolated lands into a network of increasingly connected nodes. East Asian raccoon dogs, North American mink and muskrat and South American coypu now find themselves living alongside European badgers and roe deer. Previously separated by uncrossable distances of land and ocean, these species all now live together in Europe. We have converted separate continents and distant locations within the same continent into archipelagos of partially connected regions.

This flow of species is akin to the movement of finches and mocking-birds between islands in the Galapagos. Having arrived in Europe, raccoon dogs, mink, muskrat and coypu will already be starting to take different evolutionary paths from those in their homelands, as they experience new conditions and begin to live alongside European spe-cies, which they had not previously met. Critically, while the rates at which humans are moving animals and plants are enabling species to reach many more parts of the world, they are not, in most instances, sufficient to stop them evolving separately in their new locations. There must be many millions of individuals that belong to introduced mam-mal species living wild in Europe today, but only a handful of further individuals will be deliberately released or accidentally escape each year. In other words, new arrivals from the homelands of these species will be only a drop in the genetic bucket – insufficient to stop American-origin animals becoming distinctly European. The influx is fuelling a massive increase in evolutionary diversification of the species that man-age to make these new journeys.

Of course, new connections also result in the extinction of incom-patible species, and particularly of forms that lived only on remote islands – as I have already discussed. This is a sad consequence of the increased globalization of biology. Still, extinction associated with immigration is rather rare, and increased connections between conti-nents usually result in a net increase rather than a net decrease in diversity.[14] Even on islands in the middle of the Pacific Ocean, where some of the original species died out, the total number of species on each island is far higher after contact than before. With increasing transport, each continent, island or sea is likely to hold more species in total, each with the potential to evolve into a new form. In the case of the South American deer, rats and lupins, we can see that animals and plants are liable to evolve into tens, and in some cases hundreds, of distinct species associated with different climates, habitats and geographic locations – an evolutionary radiation – when they arrive in a different continent. Once they have evolved in their new homes, the daughter species will then have the potential to spread back into their former homelands and further increase the total number of species on Earth, like a giant, worldwide Galapagos archipelago.

The biological coming together of the previously separated continents of South and North America is referred to as the 'Great American Interchange', one of the most significant biological events of the last 10 million years. It caused some species to die out, others to thrive and new species to be born. Today we are centre stage in a new Great Global Interchange – a far, far grander thing. We are watching the formation of a New Pangea, conceivably the greatest spur to evolution for a hundred or more million years. While some might think of the New Pangea as a single human-connected megacontinent, this new world is more akin to a Pangean archipelago. Each continent, and each region with a continent, and each true island, represents a node in a global network of islands. Each species that arrives in a fresh location experiences the physical conditions there, meets species it has not previously encountered, and starts to evolve into something a bit different. This new global archipelago has the potential to deliver a torrent of evolutionary changes.

Far to the north-west of the Galapagos, the evolutionary consequences of species coming into contact with each other for the first time can be seen. No longer can the human inhabitants of Kauai in the Hawaiian Islands lounge in a hammock after a tiring day in the tropical sun and enjoy the sound of male crickets – a joyful series of long and short chirps – emanating from their shrubberies. The crickets have gone silent, all because of a small fly that has travelled the world.

Kauai's crickets are themselves recent arrivals, although it is not certain whether they accompanied Polynesian colonists or stowed aboard with later merchant traders. It seems that these Pacific crickets initially set off from Australia, most likely in a series of steps in which they established new colonies on various intervening islands, and eventually journeyed as far as the Hawaiian islands, where populations can now be found on Oahu, the Big Island, and on Kauai.

Coming the other way was a fly called *Ormia ochracea* that is extremely dangerous to crickets. These flies originated in continental North America and at some point travelled westwards to Hawaii and met the Pacific crickets. Welcome to Anthropocene biology. Two species, one from Australia, another from North America, meet up in an idyllic third location. In case this sounds romantic, it isn't; nature can be a nasty business. *Ormia ochracea* likes to eat flesh – cricket flesh, to be precise. *Ormia* lay small fly larvae (rather than eggs), and the larvae proceed to burrow into their cricket victims, eating them from the inside out.

Each female fly hunts for her victims by sound. She sneaks up on the male crickets when they are engaged in their lustful songs before leaving behind her deadly grubs. The female fly has to be able to tell exactly the direction a cricket is calling from if she is to locate it, which is very difficult for an insect whose ears are too close together to be able to detect differences in the time it takes sound waves to arrive at each ear. It has overcome this by evolving an internal physical connection between its ears that is so effective, and small, that it has inspired the development of a new generation of potentially much smaller hearing aids for humans.

In the early 1990s, Marlene Zuk, a professor of biology at the University of Minnesota, noticed that the Hawaiian crickets had dropped their melodious Australian accent and that their songs had become shorter.[15] Short songs could be expected to reduce the chances that the crickets would be located by the flies, she reasoned. But the male crickets were still singing and the female flies were finding enough of them in which to leave behind their ravenous larvae. Then it all went quiet. Somewhere, a genetic mutant had arisen in the cricket population that gave the males smoother wings, and smooth-winged males do not call – it is the rubbing of the normally roughened wing surfaces that makes the sound. This could have been a disaster. Males that do not sing will find it difficult to attract females. Yet it turned out that the quiet males still had some ability to find mates, and this was less of a problem than being rasped to death by fly larvae (30 per cent of calling males at any one time harboured fly larvae).

Marlene and her fellow researchers discovered that the frequency of silent flatwings grew steadily – 90 per cent fell silent in less than twenty cricket generations. That is rapid evolution. It was far harder for the flies to find smooth-winged crickets, and so enough crickets survived in each generation for the population to recover.[16] Now, people are more likely to swing in hammocks to avoid the crickets that are crawling on the floor than to listen to their song. Whether the fly is evolving new ways to find silent crickets is not yet known, but a huge advantage is awaiting any *Ormia* fly that can home in on silent males. The contest is not yet over. The evolutionary game never is.

Exposure to flesh-eating maggots has generated differences between the Kauai crickets and those in Australia and Moorea in the South Pacific, starting the process of evolutionary separation in the global archipelago as a consequence of species meeting up for the first time. Since each species of cricket has a distinct song that enables the females to recognize their mates, this is exactly the sort of change that could eventually lead to the Hawaiian crickets becoming a new species. It could even happen on other Hawaiian islands, if the crickets develop different acoustic means of evading the flies on each island. The process of separation and evolution in partial isolation is starting already.

For the Pangean archipelago to generate increases in the number of species in the world requires the transported populations that now live in separate continents, separate habitats within each continent, and separate islands to become so different from one another that they form separate daughter species. New species are usually thought to take hundreds of thousands of years to form, if not a million, so this potential for novel diversity might seem more relevant to our descendants thirty thousand to forty thousand human generations hence than to those of us alive today.

So perhaps it is reasonable for gloom-merchants to dwell solely on the losses. On the other hand, enough examples of rapid evolution are coming to light to suggest that we have initiated a great new evolutionary acceleration: hybrid Italian sparrows have come into existence; house sparrows have evolved characteristic beak and body dimensions in different regions; the size of fruit flies and the arrangement of their chromosomes have diverged in different places to which they have been introduced;[17] and crickets have become silent. Another example is that Argentine fire ants became more aggressive once they arrived in the USA, to such an extent that the native fence lizards had to evolve new ways to escape them.[18] The question is whether any of these changes are fast enough that ex-pat populations are starting to turn into distinct species in their new homes. If so, how long will it take?

California is a hotbed of immigration, so it is an excellent place to start. Are Californians turning into new species? Foreign species are not necessarily welcomed to this part of the world, and they stand accused of ousting the natives. Among the arrivals are Spanish plants belonging to the genus *Centaurea*. California enjoys a Mediterranean-style climate, with cool and relatively moist winters and dry, hot summers, so it is not surprising that the European yellow star-thistle *Centaurea solstitialis* and its relative the sulphur star-thistle *Centaurea sulphurea* established wild populations there. The yellow star-thistle, in particular, has become so successful that it is regarded as a noxious weed – despite the fact that its spiky golden-yellow flowers supply nectar to butterflies and bees (Californian star-thistle honey is a new culinary favourite) and it mainly grows on disturbed ground where native wildflowers are rare. In any event, there is no getting rid of it now.

Long established in California, there have been plenty of generations available for the two plants to evolve in isolation from their Spanish ancestors – the sulphur star-thistle was introduced to California around 1923, allowing the Spanish and Californian populations to develop in isolation for up to eighty-six generations.[19] But could they actually have become that different after such a short period of time? No one would really have expected this to be the case, and University of Montana researchers Daniel Montesinos, Gilberto Santiago and Ray Callaway were no exceptions – ecologists and evolutionary biologists have been brought up on the 'knowledge' that it takes a very long time for new species to form. In fact, they were not thinking about it at all. The main goal of their experiment was to obtain 'pure' seeds of each population and species to use in the rest of their research. However, just to amuse himself, Montesinos, who is now at the Universidade de Coimbra in Portugal, in his own words *'playfully decided'* to transfer pollen from Spanish to Californian plants *'just to see what happened'*.

The results were very surprising.[20] Californian plants produced 44 per cent fewer seeds per flower when they were fertilized using Spanish pollen than when they were supplied with Californian pollen. Over the period since the plants were introduced to California, the compatibility of the Spanish with the Californian sulphur star-thistle has declined. Isolation in the yellow star-thistle is even greater, at around 52 per cent reduction in fertility. However, this is over a larger number of generations. The yellow star-thistle was first found growing in California in 1824, but its journey was an indirect one, via Chile, so the chances are that the Spanish and Californian yellow star-thistles last interbred 350 or so generations ago. Nonetheless, this is still exceptionally fast. The Californian and Spanish star-thistles seem to be losing the ability to mate with one another. They are on the path towards becoming separate species.

*Yellow star-thistles that are growing in California (*foreground and throughout the meadow*) are already partly incompatible with their European ancestors. They are rapidly turning into a new American species. The plants are prickly – note the spines at the base of each flower – and regarded as an invasive weed, but they also represent an important nectar and pollen source for beekeepers and for native insects.*

Because closely related species can sometimes mate with one another and produce hybrid offspring (the topic of the following chapter), the benchmark for the Californian plants to be regarded as different species is not a full 100 per cent reduction in fertility. Knowing this, Montesinos and his colleagues decided to find out what the fertility might be when you cross different wild star-thistle species with one another. They tried to fertilize yellow star-thistles with the pollen of sulphur star-thistles, and also with the pollen of yet another related species. The answer was a 65–88 per cent reduction in the number of seeds produced when crosses were made using pollen from different species. This suggests that the Californian plants, at 44 per cent and 52 per cent reduction in fertility, are probably not yet fully-fledged species, but are well on the way towards it, a mere 86 to 350 years after they separated from their Spanish ancestors. If they continue to diverge at the same rate, then they might well be quite distinct 'human-created' species within a few more centuries. Will Californians, at this point, put aside their hatred of these 'alien' plants and treat them as natives?

The arrival and establishment of star-thistles and thousands of other plants provide opportunities for insects to eat them, diseases to infect them and for birds and lizards to seek meals among their leaves and stems. As we have seen, Edith's checkerspot took to eating white-man's footprints plantains when these plants started to grow near Carson City on the eastern slopes of the Sierra Nevada range, and Taylor's checkerspot is almost entirely reliant on the new plant. There is no suggestion that this butterfly is about to become a new species; many populations have their own characteristics without turning into different species. Yet out of the many millions of populations of 'native' insects that start to eat 'alien' plants, some do change in a way that separates them from their ancestors. Then, new species start to come into existence.

The apple fly, which began to infest domestic apples in North America only a century and a half ago, is a case in point. They are charming flies, albeit not the all-time favourite of fruit farmers and those of us who like unblemished apples. Their larvae, otherwise known as maggots, or 'worms', burrow into the flesh of the apple, although they will not do you any harm if you do consume them. The adult flies are delightful, however. They have F-shaped black marks on their otherwise transparent wings, and when they perceive that they might be under attack, they twist them around and walk side-ways, waggling their wings up and down. Miraculously, this behaviour seemingly transforms them into spiders, the blackened wing-marks doubling up as pretend legs. Presumably, predatory insects or other spiders are tricked into thinking that they are dangerous, and avoid eating them.

The apple flies mate with one another on the apple trees and lay their eggs in the apples; the maggots develop inside the apples and, once they are fully grown, they drop to the ground, usually still inside the apple, then crawl out and form a pupa in the soil beneath the apple tree. Come the next generation, the pupal case bursts open, and a new adult pops out and flies off in search of the next year's developing apples. I have used the word 'apple' a lot here. Everything the apple fly does is very closely linked to apples. In contrast, their ancestors mated on wild hawthorns and laid their eggs in developing hawthorn berries, and their grubs bore into hawthorn fruits before dropping to the ground and pupating there. Hawthorn flies still exist today, eating hawthorns, much as they ever did.

Biologist Jeffrey Feder first became interested in trying to work out what was going on during his doctoral research back in the 1980s. At the time, he was working with Guy Bush, an evolutionary biologist whose research focused on the processes by which one species might become two. Jeff and his colleagues at the University of Notre Dame, in Indiana, continued this work, and they have uncovered a remarkable story.[21] The apple flies have evolved to like the smell of apples, while hawthorn flies like the smell of hawthorn. This makes sense. Each species of plant has its own characteristic odour, and the specific chemicals in these odours can be a reliable way for an insect to find its food. This keeps them apart. The apple flies, when they emerge in spring, sniff out apples and then mate on or near developing apples, while hawthorn flies go wild for the odour of hawthorns and mate on or near developing hawthorn fruits. In other words, apple flies and hawthorn flies hardly ever meet or mate with one another because they live on different trees. They have also evolved to become active at different times during the spring, because apples flower slightly earlier than hawthorns. Female apple flies need to be ready to lay their eggs

earlier in the season and this early mating further reduces the likelihood that crosses will take place between the two types of fly. By and large, they are already genetically isolated from one another, which is the principal criterion to decide whether two different animals belong to the same species or not. The apple fly is turning into a new species: it might not quite be there yet, but it is well on its way. Like the star-thistles, these flies are on the road to becoming completely separate species, in this instance in somewhere between 150 and 200 years.[22]

It does not stop there. Little flies have littler wasps inside their bodies to eat them, and the poor apple flies have three.[23] Small parasitic wasps sting the fly maggots, and the wasp grubs develop inside the bodies of the fly maggots. Grubs within maggots. Parasitic wasps often find their prey by smelling out the plants first (a whole apple tree can be smelled from further away than a maggot), before turning their attention to finding the maggots inside the developing fruits. Historically, however, the wasps would have been attracted by the smell of hawthorn. Amazingly, three quite different little wasps now find the sweet smell of apples irresistible and, even more remarkably, they are repelled by the smell of hawthorn – it sends them off in the opposite direction. These three species of apple-associated wasps also mate on or around the apples, and they emerge earlier in the year to coincide with the time when the apple-fly maggots are available for them to sting. Being attracted to apples, mating near apples, and mating earlier in the season mean that the three species of apple wasp are already becoming genetically distinct from their hawthorn wasp ancestors. Four insect species – a fly and three minute wasps – are in the process of turning into eight on a timescale of centuries, and this is all associated with just one introduced plant.

The hawthorn fly (opposite, top; on downy hawthorn) has larvae that develop inside the fruits of its North American host plants. Since domestic apple trees were introduced to North America, this fly has begun to evolve into a new species, the apple fly (middle). The strange markings on its wings impersonate the legs of a spider, and deter would-be predators. Three species of tiny wasp (bottom; Utetes canaliculatus on a native snowberry) have also made the journey onto apple, where they seek out apple fly larvae to parasitize.

There has been some debate about whether the apple flies should already be considered new species;[24] of course, there is no absolute dividing line. As two populations of one species become less and less alike, there is a continuum of divergence. Nonetheless, we have to look twice before deciding that apple flies are still the same species as their hawthorn ancestors. Most of the evolutionary biologists I have asked do not think these flies have quite crossed that line yet, in the same way that Chihuahuas and Irish wolfhounds are usually regarded as belonging to a single species. Genes can still pass from hawthorn to apple flies through occasional matings, so it is reasonable to conclude that they have not yet achieved complete reproductive isolation.[25]

On the other hand, I can't help wondering whether their opinions might be coloured by how recently apples were imported to North America. Imagine we did not know this, and then consider Jeff Feder's work again – there are differences in fly behaviour as they home in on the scents of different plants, and they eat different plants, mate on different plants, exhibit developmental and physiological differences so that they become active at the right time of year to exploit each plant and they have genetic differences associated with each plant; and then specialist wasps concentrate their attacks on apple flies. Ignorant of the historical introduction of apples to North America, we might conclude that the apple and hawthorn flies are separate species. It is the indecent speed of one species turning into two, as much as anything, that is making us reluctant to acknowledge their separation.

As species travel across the world, they are starting to diverge, establishing geographic separation from their ancestors and from other introduced populations of the same species. And resident species are already evolving in response to the influx of new arrivals, becoming separated into new ecological forms, as we have seen in the case of the flies and wasps. This geographic and ecological isolation is sufficient that these animals and plants will continue to become genetically distinct until, in the fullness of time, they become new species. Speciation seems to be unfolding in front of us at a pace that is sufficient to generate entirely new species on a timescale of a few thousand years, if not sooner.

The real avalanche of new species will, however, come tens and hundreds of millennia from now. Thousands upon thousands of plants have been introduced to new localities around the world; they already account for a third of Europe's wild-growing plants. Each one of these now-European plants is likely to start diverging from their American, Asian, African and Australasian ancestors, in a manner similar to that of the star-thistles in California. First, they will become distinct populations, then species, and, a million or more years hence, each introduced plant may form dozens of new species, associated with different habitats or geographic locations, akin to the diversification of lupins in the Andes. Animal diversity will follow. Just as checkerspot butterflies now eat introduced plantains and hawthorn flies have taken to apples in North America, European insects, diseases and fungi are all starting the process of becoming adapted to these new introduced plants. They are being joined by other plant-feeding insects, pathogens and fungi imported from distant continents, which will separate into European species in due course. And these insects will all have their own enemies. Tales like that of the hawthorn fly wasps will become the norm. Meanwhile, introduced birds and mammals are joining the throng. Stand back and a great increase in European diversity will ensue.

As humans, we have set ourselves up to keep the biological life on this planet as unchanged as possible, and part of our strategy is to repel invaders so that we can protect biodiversity 'just the way it is'. Yet our failure to control the global transport of species has a silver lining. Although the arrival of additional species speeds up ecological change (which is occasionally inconvenient), letting more species in immediately accelerates the rate of evolution and subsequently increases the rate at which new species come into existence on Earth. In other words, the biological processes of evolutionary divergence and speciation have not been broken in the Anthropocene. They have gone into overdrive. We have created a global archipelago, a species generator, which will give rise to considerably more species on Earth than existed before humans started to spill out of Africa. Come back in a million years and we might be looking at several million new species whose existence can be attributed to humans.

9

Hybrid

Yellow-flowered *Senecio eboracensis*, Yorkwort, germinated in my front garden in the summer of 2014, the first time this plant has grown in its native Yorkshire for over a decade. Gordon Eastham, the grounds maintenance manager at the University of York, also sowed it in the middle of a university campus roundabout – it is a plant that is not too bothered by human disturbance. And the City of York planted it out in one of its formal gardens for the first time since the City authorities unknowingly weeded and sprayed it out of existence a decade before. This is by now a familiar story: a species dies out, then there are attempts to bring it back.

But why does York possess a rare plant species that appears nowhere else in the world in the first place? At 54 degrees north, York is further north than Newfoundland, and at a similar latitude to the frigid Aleutian Islands, the Kamchatka Peninsula and Lake Baikal. York – not that it existed – was cowering beneath a massive block of ice a mere twenty thousand years ago and, today, much of the city is perched on top of glacial moraine, a pile of stones and sand dumped by the melting glacier at the end of the last ice age. Yorkwort can't have originated under the ice. Species usually take a million or more years to evolve, so *where* did it come from? And *when* did it spring into existence? Bizarrely, the answers to these questions are: in York, and during the last quarter of the twentieth century, before it was obliterated again by a human desire for neat, tidy pavements. Fortunately, renowned plant geneticist Richard Abbott had researched the plant's origins and had had the foresight to keep some seed.

The Yorkwort story starts two thousand kilometres away and several centuries earlier, with botanists hunting the slopes for new curiosities on Mount Etna in Sicily, in southern Italy. The botanists in question were Francesco Cupani, the first director of the botanic garden at Misilmeri in Sicily, and William Sherard, a Briton of modest origins who rose to become British Consul at Smyrna, in Turkey, from 1703 to 1716, where he established a wonderful garden and made his fortune, the latter enabling him to devote much of the remainder of his life to his true love – botany. The first step was to collect some unusual bright-yellow-flowered *Senecio*s from the middle elevations of Etna's volcanic slopes. It later transpired that these were natural hybrids between two wild species of ragwort, one of which, *Senecio aethnensis*, naturally grows high on the mountain, whereas the other, *Senecio chrysanthemifolius*, is a denizen of the lower slopes. Shipped to Oxford in the early 1700s, this mixed-up *Senecio* behaved for a while like any other self-respecting garden plant, growing where it was put in the city's botanic garden, until its wind-blown seeds fetched up elsewhere in the developing city.[1] Escaping the fertile soils of the botanic garden flower beds, it found a home in a more desolate environment – on the volcano-like walls and gravels of Oxford University's colleges and libraries. Within a century, its yellow flowers adorned the city. It acquired its name, Oxford ragwort, and the less fortunate scientific binomial *Senecio squalidis*. The plant was genetically distinct from its parents, and it did not breed with them – to all intents and purposes, it was a new species.[2]

And so things might have stayed, with one new plant added to the world's list. While the seeds of Oxford ragwort can be blown in the air from one of Oxford's dreamy spires to another, there were limits to how far they were likely to go. If they landed in the surrounding pastures and woodlands, they would be unlikely to find suitable places to grow. That is, had it not been for the opening of Oxford railway station in 1844. Industrialists, enthusiastic to transport goods and people the length and breadth of Britain, inadvertently set about constructing an Oxford ragwort conveyor. The long strips of lava-like gravels accompanying every new railway line represented a habitat to which the inhabitants of Etna were considerably better adapted than were the flowers of British meadows. Thank you, Isambard Kingdom Brunel. And then, kindness itself, the industrialists built a system of seed-suction devices – trains – which could hardly have been better designed to generate aerial vortices and move the seeds along these artificial 'lava flows', delivering them to one city after another, where they found man-made cliffscapes just waiting to be colonized. Thank you, George Stephenson. Were its history unknown, Oxford ragwort could be renamed after any other town or city connected to the railway system; it is one of Britain's most familiar flowers.

Senecio plants are not best known for their chastity. When they arrived in York, the related native common groundsel was already there. Groundsel, *Senecio vulgaris*, is less showy than the Oxford variant, lacking proper petals. When I say native, groundsel mainly grows in flower beds, building sites, unkempt pavements, yards and supermarket car parks. It is not quite clear when groundsel arrived, but it has been around for a long time. Its Anglo-Saxon name in the latter half of the first millennium was groundeswelge, meaning ground-swallow, referring to the plant's habit of invading previously bare ground. In any case, it was the *Senecio* in occupation. And when the Oxford ragwort pitched up at York railway station, introductions and a little bit more were made. A new hybrid came into existence in 1979, give or take a few years, and the hybrids bred true.

The ability of the parents to cross at all was quite impressive, because the groundsel had forty chromosomes – the structures in every cell that host each species' genetic code – whereas the Oxford ragwort only had twenty. The York hybrid version opted for forty, perhaps explaining why it can interbreed more readily with its groundsel parent than with the incomer from Oxford. Nonetheless, the York plants showed the capacity to establish a genetically distinct and self-perpetuating population in the 1980s and 1990s. Richard Abbott was happy to recognize the York plants as a new species, which seems reasonable, given that many other plant species are also a bit promiscuous.[3] So, *Senecio eboracensis* was born, Eboracum being the Roman name for York, which itself refers back to Iburakon, meaning place of the yews in a former Celtic language. Species number two. By the 1990s, *eboracensis* could be found growing in some profusion between the cracks of the pavements on the banks of the River Ouse, a short seed-blow from York's railway station. However, it disappeared again when the City of York authorities decided to clean up its pavements. Minus one species. Or, it would have been but for the collection of seeds that Abbott and his colleagues still had, and donated so that it could again be grown in the city of its origin.

So, we are still at plus two species: Oxford ragwort and Yorkwort. And it did not stop there: Oxford ragwort's tour of Britain has been extensive, hybridizing with common groundsel elsewhere, often spawning a plant called *Senecio baxteri*. Because thirty-chromosome *baxteri* is sterile, that doesn't count as a new species. However, a few individual *baxteri* plants must, by chance, have experienced a developmental 'error' somewhere in North Wales, and a fertile sixty-chromosome version was born. Hence *Senecio cambrensis*, Welsh groundsel, arrived on

Yorkwort, Senecio eboracensis, *a new hybrid species that came into existence in the City of York at the end of the 1970s. Yorkwort formed as a consequence of humans moving its ancestors to Britain and providing suitable disturbed and stony ground for the plants to grow in.*

the scene[4] – speciation in an instant. Species number three. The same thing happened in Edinburgh, but they were very like *cambrensis* and died out again, so perhaps we should not add the Edinburgh hybrids to the credits.

Three new species and an array of other hybrid forms have all been generated in three hundred years as a direct consequence of human actions: transporting a plant across the world from Sicily to Britain, building artificial cliffs and beds of gravel, creating trains that could move the seeds great distances and thereby allowing Oxford ragwort to meet up with groundsel, which was itself thriving in human-created habitats. This spawning of around one new species per century is about ten thousand times faster than the conventional rate at which new species are expected to form. And human actions are responsible.

Are these *Senecio* plants a one-off, or are we initiating a mass diversification of new species on Earth, at the same time as others are being driven extinct? Californian weeds are diverging from their European ancestors, apple flies have come into existence, wasps that kill apple flies are thriving, Italian sparrows formed when house sparrows hooked up with Spanish sparrows, and now *Senecio* plants are turning into new species at immodest speed. *Senecio* hybrids are likely to be spawned wherever Oxford ragworts and their relatives meet up with groundsel and other *Senecio*s throughout Europe. Give it a few millennia and each major European city might have acquired its own unique type of *Senecio*, many of which could be distinct enough to be regarded as a separate species.

As far as we know, no British plant species has become globally extinct over the same period that the *Senecio* species were diversifying. Some species have died out in Britain, but still survive elsewhere, in continental Europe. This means that the contribution of human actions within the geographic confines of Britain over the last three centuries has been to increase the number of plant species on the planet. The new hybrid *Senecio*s are nowhere near as distinctive as some of the world's extinct species, however, so they should not be regarded as like-for-like replacements. The great auk, for example, was a flightless seabird that used to nest on offshore islands surrounding the British mainland and on other rocky islets in the North Atlantic; it became globally extinct in the middle of the nineteenth century because we killed them all for their meat, eggs and feathers. It was unique. The great auk's ancestors split from those of the razorbill some 25 million years ago (the razorbill is a smaller and fly-ing version of the great auk, and survives today – just as we saw in New Zealand that the flying pukeko survived while the pedestrian takahe is flirting with extinction) and so it was genetically one of a kind. The different *Senecio* species are genetically much more similar to one another than the great auk was to the razorbill.

On the other hand, new *Senecio* species are living organisms. The Anthropocene epoch is indeed the regrettable end of the evolutionary story for many animals and plants, but it also represents the begin-nings of many new stories. Evolutionary acceleration is starting to gather pace.

Britain is perhaps the only part of the world where we can begin to estimate how fast new species are being added to the world list, given our long interest in gardening and natural history, as well as it being the birthplace of evolutionary biology. Two different species of monkey-flower, one from South America and one from North Amer-ica, escaped from our gardens and, on two separate occasions, their hybrid offspring have doubled their chromosomes and turned into a new species that now lives along Scottish stream-sides. This presents a real dilemma for conservationists – should they kill the new alien plant or celebrate its arrival on Earth? A marsh-dwelling cord grass originated when North American *Spartina* grasses hybridized with British *Spartina* on the intertidal expanses of mud close to South-ampton. A new hybrid species was born, *Spartina anglica*, proudly bearing its country of origin in its name, before spreading around the coastline of Britain, and then setting off to create its own English cord-grass empire in estuaries across the world.[5]

In a more domestic setting, we have the Kew primrose, which first came into existence in 1898 after plant hunters imported one of its parents from the Himalayas, and the other from North Africa or Arabia. Together at last in the Royal Botanic Gardens at Kew on the south bank of the River Thames in London, pollen could meet stigma. And their offspring, the Kew primrose, was born.

All told, seven or eight new plant species have come into existence in Britain since 1700. Two to three species per century in Great Brit-ain might not seem like much, but this rate of formation of new species is about a hundred times faster than new plants have come into existence in the geological past.[6] If the same rate of hybrid plant speciation is replicated over the rest of the world's land surface (which may not be true yet but could well be the case within the next century, given the acceleration in transport of species between the continents), then – very roughly – a thousand new plant species will be generated per century.[7]

The formation of new species is only the tip of the hybrid iceberg. Horticulturalists have brought tens of thousands of varieties into existence by making crosses between different garden plants. And hybrids are not confined to gardens. Recently published maps identify over nine hundred different crosses between plant species living in the British countryside.[8] Many of them will simply hang around for a while and then die out – but some are here to stay. Purple-flowered rhododendrons, which beautify or ruin our British landscapes (depending on your perspective), are mainly of European origin, but they contain a hint of genes from two North American rhododendron species.[9] Their 'pure' European ancestors were apparently somewhat tricky to culture, which is not surprising, given that they come from much warmer regions in Spain and Portugal, whereas the new garden hybrids were able to cope far more effectively with the cold weather. The vast majority of the genes in the new hybrids still come from the European species, but it has been suggested that the 'American genes' may make the soft Europeans a little tougher (winters have become milder since the 1800s, too, which may have helped). The resulting hybrids are extremely successful, forming evergreen under-storeys to deciduous woodlands on acidic soils and spreading out to form low-growing shrublands over heaths, moors and dunes. Traditionally known by its European botanical name, *Rhododendron ponticum*, the wild-growing rhododendron thickets are now referred to as belonging to *Rhododendron x superponticum*, the *x* denoting its hybrid origin and the *'super'* providing a reminder that we should perhaps disapprove of its ability to romp across the countryside. Recent headlines describe the *'threat from aggressive rhododendron'* in England, a *'30 years war'* against the plant in Ireland, and the *'rhododendron menace'* in Scotland; but Wales apparently has the answer, as reported by the *Welsh Daily Post*: *'Meet Snowdonia's "Rhodocop": Gruffydd is looking to root out alien invaders*.' While hybridization has not instantaneously generated a new species, in this case it has formed a distinct population that has the potential to evolve into a new kind of rhododendron over a much longer period of time. It is a botanical addition to the world, however much Gruffydd and his friends love to hate it.

It is not something about the air in Britain that induces plants to hybridize and form new varieties and new species – although the British obsession with gardening has helped. The same phenomenon is under way elsewhere. Moreover, it has been taking place ever since people started to cultivate plants. Our ancestors created new species of wheat by propagating large-seeded varieties of hybrid wild grasses in the Middle East,[10] they brought us hybrid peanuts that contain the chromosomes of two wild relatives from South America,[11] and they generated three new hybrid cabbage species in Eurasia and Africa.[12] The United Nations Food and Agriculture Organization lists at least six human-created species in their inventory of the world's most important sources of human food around the world. And these are just the completely new species. Most of the other major crops also contain at least some genes from more than one originally wild plant species. GM purists beware – every meal you eat is likely to contain a mixture of genes that did not exist before humans came along!

Away from agriculture, golden yellow and purple salsify plants from Europe were brought together in North America, and the hybrids that formed between them experienced genetic changes. They have become two or three entirely new species that live in the US states of Washington and Idaho.[13] And new hybrids between European hawkweeds have been spawned where they have been introduced to New Zealand.[14] The current rate at which new species are forming on Earth is starting to look as though it is the highest ever, or at least the highest since animals and plants first colonized the land.

The formation of one new creature from two is not new. Nearly every cell in our living bodies contains energy-giving structures called mitochondria, which originated as free-living microbes in a primeval world that existed before anything that we recognize today as 'animal' or 'plant' came into existence. Similarly, the green of plants comes from their chloroplasts, the descendants of ancient microbes that still live inside their leaves and stems, billions of years after two separate organisms first started to live together. This historical merger allows plants to capture the energy of sunlight and turn carbon dioxide from the atmosphere into the sugars they use to grow. The abilities of cows and plant-sucking aphids – and humans – to digest these plants also depends on the existence of different microbes within the guts and bodies of animals. In short, every single animal and plant contains a mixture of genes that originated from different species. However, these biological joint ventures with microbes are rather different from the hybridization between two closely related species of *Senecio*. It is these hybridization events that appear to be on the increase.

When modern humans first spilled out of Africa, we were not the first humans in residence in Europe and Asia. Neanderthals were living in Europe and Denisovans in Asia. And one thing led to another. We may never be sure whether it was rape during acts of war, abduction and enslavement, orphans reared as children of the other species, or loving, harmonious relationships. Knowing humans, it was probably all four. Whatever the details, we mated, produced some fertile offspring, and the genes of both Neanderthals and Denisovans live on in us today.

Thanks to the ability of molecular biologists to read our genes and extract DNA from ancient Neanderthal bones, we know that roughly one to a few per cent of each person who lives outside Africa comes from our Neanderthal ancestor genes (apart from more recent migrants from Africa, who contain only trace levels of Neanderthal). Perhaps even more remarkable, each of us contains slightly different genes that were derived from Neanderthals, such that 60 per cent or more of the genome from Neanderthals still lives on in the modern human species.[15] So Neanderthals are less than half extinct after all – they are us! In Asia, even more Denisovan genes mixed with those of our Ethiopian-origin ancestors as they spread into eastern Asia, Australia, the Pacific and the Americas.[16]

Today, the mixing continues. Consider New Zealand biologist Jacqueline Beggs, whose ancestors include Europeans who bear some of the genes of Neanderthals, and Maoris who bear some of the genes of Denisovans. It does not stop there. The split of the human and chimpanzee lineages long ago seemingly took place over several million years (different chromosomes diverged at different times), which implies that hybridization between closely related ape species happened then, too. If we add in her mitochondria, and the trillions of microbes living in her body, then Jacqueline is a multiple-species and multiple-hybrid animal, as we all are.

Some might argue that the first modern humans, Denisovans and Neanderthals were just three branches of one human species rather than separate species, although there is genetic evidence that male

Jacqueline Beggs has European ancestry, which includes hybridization between modern humans and Neanderthals, and Maori ancestry, which includes modern human x Neanderthal crosses with Denisovans. She and all humans also show signs of past hybridization events between a range of ape species that lived millions of years ago. Jacqueline is holding a New Zealand kakapo, the world's heaviest and only flightless parrot. It is also nocturnal and breeds only when several species of tree produce heavy fruit crops. It is incompatible with introduced mammalian predators, but intensive captive breeding and releases on to predator-free islands – here, on Codfish Island – have brought numbers back to 125 adults and 33 first-year chicks (in 2016).

hybrids between modern humans and Neanderthals had reduced fertility. And perhaps different plants that can still hybridize are not fully separated species. As we have seen when discussing whether Chihuahuas and wolves, or apple flies and hawthorn flies, are separate species, it is often tricky to draw a clear dividing line, because speciation represents a continuum of separation. In any event, the overall story is unaltered by such debates. Whichever view one takes, human ancestry has not been an ever dividing tree of life but a reconnecting thicket that comes back together every so often. When hybridization takes place, whether between distinct populations or closely related species, the hybrids then move on with a mixture of genes. Mixing continues. Today, modern Africans are again interbreeding with people whose ancestors have spent up to a hundred thousand years living outside Africa, and whose Denisovan and Neanderthal genes have been living outside Africa for about half a million years. Come back in a thousand years and there is a good chance that every human on the planet will contain at least some Denisovan and Neanderthal genes, as well as the genes of an unknown and extinct African species that has only ever been detected in the genetic code of present-day Africans. And this genetic diversity is likely to be beneficial to the long-term future of humanity.

The technologies to read and interpret our genes are still in their infancy, and the only reason we know so much about the hybrid origin of people is because we are people. We have stared into our own genetic mirror more deeply than into the DNA sequences of other species. Yet, from what we already know, it seems that the human story is not unusual. Different kinds of animals and plants do sometimes mate with one another, and they do sometimes produce fertile hybrid offspring. Those fertile hybrids may then either integrate back into one of the parent species (taking some genes from the other species back with them, as in humans and rhododendrons) or go on to produce a separate species that spreads out across the world, like the hybrid English *Spartina* cord grass that has colonized North America, Asia and Australia in modern times. But this understanding of the importance of hybridization is very recent.

The more people look, the more it seems that hybridization is the norm. South American *Heliconius* butterflies have acquired wing colour genes by mating with one another, and this process has generated a new hybrid species in valleys high on the eastern flanks of the Andes.[17] Red wolves from the south-eastern United States appear to be about 80 per cent coyote and 20 per cent wolf.[18] Appalachian tiger swallowtail butterflies share the genes of two parent species.[19] The Atlantic's Clymene dolphin is a self-perpetuating hybrid between spinner dolphins and striped dolphins.[20] Alaskan grizzlies have polar bear genes in every cell of their bodies.[21] Hybridization is really quite normal, and always has been.

Ever since Darwin, we have represented the evolution of life as an ever dividing, branching tree, but this is not quite how it works. The idea of a tree of life should be replaced by the image of a more tangled mosaic of interacting lives in which closely related species, in particular, may continue to exchange some genes for millions of years after they first separate.[22] The consequence is that different genes in the bodies of each one of us arrived there by slightly different routes. This is absolutely the norm for bacteria, in which 'species' that apparently diverged early in the history of life still exchange genes. The reality that life is a tangled thicket demands a change in attitude towards recently hybridized animals and plants. They are not damaging the tree of life – this is how the tree of life grows. We think no more or less of any other human, I hope, on account of the fraction of their genetic code that is derived from Neanderthals, Denisovans, Ethiopians, or any other lineages of humans. It is not relevant to us. My own inner Neanderthal does not alter my humanity, and anyone who nurtures racist thoughts should contemplate the reality that their own bodies contain the genes of more than one former human species, and the genes of many different pre-human species, too. So it is with plants and animals that have hybrid origins.

There are two main reasons why humans are causing such a rapid increase in the rate at which new hybrids are forming. One is the new opportunities that we provide: an artificial canal dug by engineers enabled two types of sculpin fish to meet and mate with one another in the Rhine river system in Europe; coyotes carry dog genes around rural communities in New England; and house sparrows followed human cultivators out of India, met Spanish sparrows and spawned the hybrid Italian sparrow. New habitats and connections represent opportunities for species that were previously separated to meet up and hybridize and, in some cases, turn into new species.

One such example is the *Lonicera* fly. In 1997, Pennsylvania State University student Dietmar Schwarz was out for a jog when he stopped off to take a look at a honeysuckle bush, which was of particular interest to him because he had previously studied insects that eat honeysuckles, back home in Germany. Dietmar spotted some maggots, collected them and so started an accidental research project. The honeysuckles growing in the area included hybrids between creamy-flowered *Lonicera morrowii*, a honeysuckle from Japan, Korea and parts of China, rose-flowered *Lonicera tatarica* from Siberia and pink-flowered *Lonicera korolkowii* from the dry mountains of Afghanistan and Pakistan. As in so many other cases, these re-united distant relatives had started to exchange genes, in this instance within the 250 years since their introduction. While the foreign plants and their hybrids might be defined as invasive noxious weeds by the human inhabitants of North America, the local flies were not to be put off. To them, the honeysuckles represented a new potential habitat. The insects in question were the North American blueberry fly, whose maggots eat North American blueberries, and the North American snowberry fly, the maggots of which like nothing better than to chew their way through North American snowberries. However, while blueberry flies dislike snowberries, and snowberry flies eschew blueberries (and so they do not normally meet or interbreed), both these native insects were prepared to check out the berries of introduced honeysuckles.[23] It seems likely that this is where the two species of fly met up, and mated – genetic analysis reveals that the honeysuckle maggots are hybrids between blueberry and snowberry flies.[24] The love child of the blueberry and snowberry flies had a liking for honeysuckle berries, and the *Lonicera* fly dynasty has not looked back since. Dietmar's convenient rest stop resulted in him discovering a new hybrid species: new plant, new opportunity, new insect species.

The second and far more pervasive reason why hybridization is on the rise is because humans are moving so many species around the world directly, bringing them into contact with their distant relatives.

In Scotland, red deer are starting to hybridize with imported Asian sika deer, especially on the misty Kintyre Peninsula, which is a long, thin promontory that rudely protrudes from the west coast of Scotland.[25] In 1893, eleven sika deer from the southern Japanese island of Kyushu were released into the Carradale Estate, but it didn't take long for them to escape. The sikas established a feral population and spread up the peninsula towards the Highlands, where they encountered larger numbers of red deer. Their ability to hybridize may seem surprising because the Asian and Scottish deer last shared a common ancestor about 6 million years ago, which is not so different from the time that humans separated from chimpanzees. Nonetheless, up to one in five hundred matings are thought to be between the two species of deer, rather than within their own kind, and this has been sufficient to enable their genes to mix. Given this level of genetic mixing after 120 years, it seems likely that most individuals will eventually bear at least some genes that originated in each species.

This hybridization has led to much hand-wringing among the conservation community. As the Great Britain 'Non-native Species Secretariat' sika deer factsheet puts it: *'Hybrids with the native . . . red deer are fertile, and . . . hybridization . . . is threatening the genetic integrity of both red and sika deer.'*[26] This is interesting wording because it reflects the old thinking that the tree of life should have a perfect branching structure. Individuals and populations that contain a mixture of genes that originate in different species are regarded as worse than either of their 'pure' parents. Although some of the hybrids can be identified by their physical appearance, gamekeepers are not able to tell many of them apart from the parent species, so removing all sika genes from the red deer population is not much more practical than the prospect of removing our own Neanderthal genes. The reality is that a new Scottish and British population is being established that will contain a mixture of red deer and sika genes, as well as a touch of American wapiti; and this new population will likely evolve in new directions. I find it difficult to understand why this should be regarded as a threat.

Similar anguish has been expressed in North America, although the details of the 'beefalo' story are rather different. It all started as a mishap back in the 1700s, when the odd mating took place between

North American bison and domestic cattle, which are themselves descendants of wild European aurochs. These hybrids did not work out particularly well, and so it was not until the 1880s that serious attempts were made to produce hybrids between the two. The idea was to generate animals that were more docile than bison but hardier than cattle, which ranchers could run on rangelands where the wild bison had been hunted to extinction. These attempts were not particularly successful either (until the 1960s), but the wild plains bison population was down to a few hundred animals in the late nineteenth and early twentieth century, such that even a modest amount of hybridization was likely to leave its mark. The consequence is that most plains bison herds now contain at least a smidgen of cattle genes.[27]

One such population is happily munching its way through meadows on the north side of the Grand Canyon, where they are blamed for *'destroying water sources, vegetation, soil and archaeological sites'*.[28] A debate has been raging as to whether they should be culled because of their impurity, spawning all sorts of panic-stricken headlines like: *'Scheming buffalo herd roams amok at Grand Canyon'*, *'How do you solve a problem like the "beefalo"?'*, *'A beefalo invasion is causing trouble in the Grand Canyon'*, *'Failed experiment beefalo "destroying Grand Canyon" with uncouth ways'*, *'Grand Canyon: Bison hybrids trampling . . . sacred sites'*, and, it being the US, *'Bison problem? Let Arizona hunters deal with it'*. The official response agreed by the US National Park Service is to cull the 400–600 strong herd down to 80–200 animals, and to 'improve' their genetics, which may involve releasing 'pure' bison females into the herd.[29] Having lots of bison-sized animals in the landscape will certainty alter the vegetation, but it seems a tad unfair to blame the changes on the hybrid nature of the animal. The animals look like bison, and it is the bison genes that enable them to thrive in the harsh environments which they stand accused of damaging.

Furthermore, it seems completely unreasonable to suggest that today's half-million North American bison (with a genetic soupçon of auroch) can possibly be having anything like as much impact as the 30 to 60 million that used to roam North America as recently as the 1700s. An additional irony is that it has just been discovered that the European bison, the wisent, originated as a hybrid between wild auroch cattle and steppe bison (a close relative of the American bison) about 120,000 years ago.[30] Intensive conservation efforts are lavished on the hybrid wisent. While conservationists will dismiss this hybridization as too long ago to worry about, measured against the long timeline of life on Earth, this 'ancient' hybridization happened at more or less the same time as the modern hybridization between the cattle and American bison. Yet our responses to the two events are very different.

There seems to be a deep-seated feeling that every animal should be pure, whereas we know that, in reality, every individual of every species is likely to have at least some degree of past hybridization lurking within its genes. This attitude to genetic purity makes no biological sense, and it is not a viable way of managing the Earth. The exchange of genetic information between organisms that we usually think of as distinct – hybridization – is one of the ways in which new genetic forms come into existence and genetic fitness can potentially be increased. Hence, it is an important part of the history, present and future of life on Earth. The idea that humans can or should police hybridization is ludicrous.

These new connections are unlike any previous period in the Earth's history. The explosion that killed off the dinosaurs, and the other four major episodes of mass extinction in the last half-billion years, did not transport vast numbers of the survivors around the planet on a timescale of hundreds to thousands of years. There was no equivalent bringing together of species from different regions. The nearest equivalent is when continents have collided, generating periods of increased extinction and increased diversification. However, the biological joining of South and North America took place over several million years, rather than centuries, and the coming together of Pangea from previously separated continents took about 100 million years. Even when the world's land did form a single continent, the connections were not as great as now. Pangea was the combined size of all today's continents, so movement between distant locations would have been limited – just as many species have not (until recently) been able to move between the opposite ends of Eurasia, between the east and west coasts of North America, or between eastern and western Australia. We are even transporting species to remote oceanic islands, which would have remained isolated when Pangea existed. Given the geographic distances and rapidity of connections that are taking place in the Anthropocene, there is no precedent since multi-cellular life forms colonized the land.[31] I find it difficult to imagine a period in the entire history of terrestrial life on Earth when the speed of origination of new evolutionary lineages could have been faster, as a result of the combined forces of populations arriving in new locations and starting to diverge there, the previous residents becoming adapted to the new species that arrive, and new hybrids coming into existence as species meet up for the first time in new habitats and new geographic locations.

This brings us back to the rate at which new species are forming. More new plant species have come into hybrid existence in Britain in the last three hundred years than are listed as having died out in the whole of Europe, and the one casualty is a species of violet that is closely related to other species of European violet.[32] I'm sorry to see it go, of course. The number of new plants that have formed as novel hybrid species in North America, such as the American salsifys, is also greater than the number of species of plant that are recognized as having become extinct. The beautiful creamy, camellia-like flowers of the Franklin tree, whose pendulous leaves turn a fiery red in the autumn, disappeared from the banks of the Altamaha River in the state of Georgia in the early 1800s. This appears to be the only higher plant listed by the International Union for the Conservation of Nature (IUCN) as having become extinct in the wild on the United States mainland. This is a real shame, but you can still buy its seed, obtain cuttings and order potted plants. It is not actually extinct. It has abandoned the Altamaha River in favour of its new habitat – suburban gardens and parks. The IUCN list is probably not complete[33] but, on the face of it, the current rate at which new flowering plant species are forming in the landmass of North America is at least on a par with the extinction rate, as in Europe.

Given that we lack a proper inventory of all the new species that have come into existence (just as the inventories of living and extinct species are incomplete), we should be cautious and simply conclude that plant originations and extinctions may be comparable in these two parts of the world. I do not wish to downplay the losses, particularly in regions where major new land clearances are taking place. The world has probably lost more plant species in the last three hundred years than we have gained, and more vertebrate animal species have disappeared than new ones have formed. Yet new hybrid animals as well as plants are coming into existence and populating the Earth at a faster rate than ever before.

The human era is undoubtedly a time of unusually rapid extinction. We should regret the losses – but we should also applaud the gains. We are living through a period of the rapid formation of new populations, races and species. In the end, the Anthropocene biological revolution will almost certainly represent the sixth mass genesis of new biological diversity. It could be the fastest acceleration of evolutionary diversification in the last half-billion years. Some might discount these new species as weeds and pests, but that is a reflection of the human mind, not a fundamental attribute of these new forms of life. All forms of life simply come into existence on account of their individual histories and take advantage of the resources that are available to them. If some of these thrive at our expense, that is our problem, not theirs.

PART IV

Anthropocene Park

Prelude

It is perhaps time to take stock of the changes that are taking place, before we move into the final section of this book, which concentrates on our attitudes to the biological world and our strategies to protect it.

Parts I and II illustrated the dynamic nature of wildlife. Nature never sits still. While some species have declined or become extinct, others have thrived, and nearly all the species that most of us see around us every day are, at least to some extent, beneficiaries of a human-altered planet. Our encounters with large animals have left many species extinct, but our domestic animals and plants are now widespread, and the largest wild species that have survived are starting to recover in many regions of the world. While it is inevitable that fewer species now live in any given square metre of fields that are used for intensive food production, the world's biota has been remarkably resilient. We altered the world's habitats, and vast numbers of species have taken advantage of the new conditions. We changed the atmosphere, and hence climate, and species have spread into new regions as a consequence. And we have directly transported species to new parts of the world. The upshot is that all species are now living in human-altered environments and a high proportion of all the world's species are also living – at least somewhere – in new locations. These represent biological gains for animals, plants, fungi, microbes, viruses and any other kind of organism you wish to mention. Furthermore, the overall consequence of the arrival of new species into each region, be that a particular country or island, has been to raise the biological diversity of that part of our planet.

These dynamic changes are not fundamentally different from those in the past, other than that humans are directly or indirectly responsible for them. Dynamic changes to the locations where species live is how they survived the ice ages. It is how species are surviving now and, if we wish to maintain the world's biological resources, which I will argue in Part IV is a sensible human strategy when we face an unknown future, we should not try to halt biological change. I would certainly advocate that we tackle the underlying causes of change: we should stabilize and then reduce the human population, minimize levels of harmful consumption, obtain our food in ways that reduce our footprint on the Earth, and minimize and when possible recycle the waste we produce. We should also reduce our greenhouse gas emissions. But we should not normally attempt to halt how the biological world responds to the consequences of humanity, except when those responses are directly and obviously injurious (we will, of course, want to control new human diseases, crop pests and pathogens that afflict our livestock). Biological change is how life on Earth survives. Treating each arrival of a species in a new habitat or geographic location as something to be resisted will, in most instances, result in failure, and it is ultimately counterproductive. Humans must adapt to and help direct change, rather than attempt to preserve the world in aspic.

Over longer durations, the entire history of life is a narrative of change, involving both changes to the distributions of different types of organisms, and evolutionary changes to the sorts of creatures that live on Earth, as I discussed in Part III. Life is the product of evolution, and evolution is how life meets each new challenge. When the world is altered, as it has been by the rise of humans, it is met by evolutionary change. Some species do not make it, but others continue to live, and go on to perform roles in the next act of the world's evolutionary play. This evolutionary changing of the guard involves certain types of species surviving while others (such as flightless island birds) disappear. It also involves evolution within every population of every species, evolution when species arrive in new locations, and hybridization when dissimilar but related forms meet. These processes of biological renewal have accelerated to unprecedented levels as humans have transformed habitats and transported species around the world. A massive evolutionary genesis has been taking place over the last ten thousand or more years, and the rate of evolutionary change continues to accelerate. Life on Earth survives because it changes.

The inevitability of ecological and evolutionary change in the human epoch brings us back to our starting point. The fundamental process of life is simply the passage of information, chemically coded in our DNA, and this information is used to build the bodies of each successive generation. The consequence of the error-strewn transfer of information through time has been an immense diversification of life on Earth, in which there are both gains (new types come into existence) and losses (some types disappear). This dispassionate process means that the only value judgements we can make about these gains and losses are our own, as individual humans and as a society. As such, we need to contemplate carefully the relationship between humanity and nature. This is the subject of Chapter 10, in which I argue we should genuinely accept that we are 'as one' with nature. We are part of nature. Once we have sorted this out, it is possible to take a considerably more positive attitude to natural changes and to conservation – the topic of Chapter 11. We can think of the world as a biological park, Anthropocene Park, with ourselves both as custodians and inmates. For better or for ill, this is the world we inhabit.

10

The New Natural

Oscar Wilde prided himself on being as unconventional as possible, but when he dismissed nature as '*a place where birds fly around uncooked*' and explained that '*if nature had been comfortable, mankind would never have invented architecture*', he was simply expressing the prevailing opinion of the nineteenth century. He enjoyed taunting urbanites who retreated to the countryside at the weekend; nonetheless, they shared his views about the separation of nature and humanity. The only difference was that Wilde wanted to escape from nature, while they wanted to escape to it. Looking past his witticisms, Wilde was, on this occasion, giving voice to a societal norm. The distinction between nature and humanity was a God-given separation for the majority of the population who believed in a deity, and it was still widely accepted by those who did not.

The separation myth persists to the present day,[1] despite the fact that Darwin and Wallace published their theory of evolution while Wilde was still a child, over 150 years ago.[2] For example, the English language requires me to refer to 'who' when referring to a human, but 'which' or 'that' when speaking about animals and plants, as though they were stones. Adventure programmes and wildlife documentaries frequently serve up a diet of an idealized and separated nature.[3] Nature conservation is commonly represented as an activity whereby humans do 'good things' for nature, while others take a more utilitarian view, in which nature is something 'out there' that benefits humans. And scientists usually treat human impacts as external drivers of change rather than as integral parts of the system. The separation myth permeates scientific and journalistic writing to such an extent that our planet's natural history is even said to have become its unnatural history.[4] Yet we know, objectively, that the human species evolved naturally, so humans must be natural. We are part of nature. Accepting this, the perspective that 'humans are making nature less natural' is equivalent to saying that 'nature is making nature less natural'. This does not make sense.

Consider the past. The catastrophe that befell *T. rex* and the other huge dinosaurs was unusual in that it originated in space, whereas most other mass extinctions in the history of the Earth appear to have been caused by natural internal changes within the Earth system itself – some geological and some biological. Outpourings of lava, climate-altering changes to the position of the continents and depletion of oxygen in the oceans have caused mass extinctions whenever these cataclysmic geological events[5] created conditions that were beyond the tolerances of the majority of species then alive. Other extinction events have arisen when entirely new types of living thing have evolved. The evolution over 2 billion years ago of photosynthetic Cyanobacteria that could trap energy from the sun ultimately generated so much free oxygen – a waste product of the new chemical reaction – that they made the atmosphere and oceans toxic to most of the Earth's previous inhabitants. That biological event opened up new evolutionary opportunities for multicellular life forms that could develop an oxygen-based metabolism. Crickets, crabs, cuttlefish, cod, caiman, crows and cheetahs would not be possible but for that oxygen.

Today, a rapid increase in the rate of extinction is again being driven by an unprecedented evolutionary event arising within the Earth system: the rise of an unusually brainy and linguistically capable primate. Some 7 million years ago, our predecessors were African primates.[6] They were intelligent, sociable apes who lived predominantly on the ground but climbed trees for food and used a small number of basic tools. If we saw a living specimen of this animal, we would be unanimous that it was not human and that it was part of the natural world. But it continued to evolve. We can surmise that two or more populations of this ape separated, living in geographic regions within Africa where the climate, vegetation and availability of food differed. These populations would have diverged, such that the apes from each region would no longer possess exactly the same behaviours and physical attributes, in the same way that humans from different parts of the world can today be recognized. Given enough time, these separate populations eventually split into two or more related species, although they continued to hybridize every so often. But still we would agree that they were not human and that they were part of the natural biological system.

Many new apes emerged along the way, of which just two branches survive today, one becoming ground-living humans, the other forest-adapted chimpanzees and bonobos (our shared ancestry with the other great apes goes back to an even earlier time), which evolved elongated fingers and an increased ability to swing through the trees.[7] Our own line came out of the trees, ran on its back legs, lost body hair, lived in complex social groups, enlarged its brain, developed more sophisticated tools, controlled fire and acquired complex communication. These evolutionary and cultural developments in turn led to the runaway social innovations of the last few thousand years, including our ability to use intricate languages, domesticate animals and plants and build cities, as well as our capacity to develop religions, dictatorships, monarchies, democracies, bureaucracies, trade networks and other systems of co-operation and control on increasingly large scales.

When we contemplate the biology and impacts of humans on the Earth, there is no doubting that *Homo sapiens* is an extremely unusual animal. But at what point in the unbroken sequence of generations should we decide that humans ceased to be part of nature, and at what point did the effects of humans on the rest of the world become unnatural? There is no scientific or philosophical justification that could be used to separate this continuum of ape-then-human animals into two qualitatively separate categories. Evolution took place for long enough that we are recognizably human, and chimps are recognizably chimps. In many ways, we are still very similar, however – we shared the same great-great (add something like 250,000 more 'great's) grandparents. Chimps can also go grey and show receding hairlines as they age, they live in complex social groups, sometimes run on their back legs, have extremely large brains and use tools. They can communicate sufficiently to pass on cultural information, undertake group hunts and organize the equivalent of warfare between chimp communities.

More recently, my own modern human ancestors mated with Neanderthals, evolved the ability to digest milk as adults and developed pinkish skin. Pale skin enabled my female predecessors to produce sufficient vitamin D to support the growth of their children during pregnancy and to lactate in the dark northern winters. The European population was gradually growing apart from the ancestral Ethiopian population from which we originated. The differences that we see between human populations that live outside Africa have arisen within the last 60,000 to 100,000 years – or 2,500 to 4,000 generations (1–2 per cent of the separation between us and chimps) – as a result of new genetic mutations, natural selection which favours some

A young orangutan in Sabah, Borneo. Both humans and orangutans have evolved naturally in the half a million generations since we shared the same parents, hence the impacts of humans and orangutans on other species are natural.

of these mutations over others, continued migration and some degree of hybridization with other types of human who already inhabited the Eurasian continent. The usual natural evolutionary processes were taking place. There is a possibility that, perhaps a million years from now, we might have ended up with many different human species spread across the world, just as a population of an earlier *Homo* evolved into a distinctive dwarf species on the Indonesian island of Flores, where it survived until its extermination (perhaps by *Homo sapiens*) about fifty thousand years ago. This separation into many different human species could have been our destiny, had it not been for the torrent of human movement around the world that we have seen in recent times, the consequence of which is that the world's human genes are ending up back in one big Pangean melting pot. Humanity is setting off on a new evolutionary journey.

There is absolute scientific certainty that we evolved from apes (we still are apes), that we are still evolving and that the evolution of humans was a perfectly natural event in the history of life on Earth. We were then able to overcome most diseases that would otherwise limit our population, engineer the environment so that we could survive across most of the land surface, and commandeer more and more of the world's resources, as well as bring down mammoths and the rest of the great beasts that we drove to extinction. Extensive agricultural landscapes and the architecture and cities that Oscar Wilde approved of exist because humans evolved, and hence permitted the success of sparrows, Oxford ragwort and the other species that live in them. Entirely natural. The new distributions of species we have transported across the world are equally natural. All these things represent an indirect product of evolution.

The very fact that we continue to conceptualize a separation between humanity and nature over a century and half after Darwin and Wallace developed their 'dangerous idea' implies that there must be something quite fundamental driving the sense of 'other' whenever a human contemplates nature. And the most fundamental thing about us is that we evolved. Evolution by natural selection brought us into existence in the first place, and it is evolution that has also made it difficult for us to accept that we are simply a part of the natural world. Evolution has programmed us, by means of a complex system of biochemical, electrical and genetic mechanisms, to love and protect our offspring, mates and wider family, because their safety and subsequent reproduction is the primary means by which our genes are transmitted to the next generation.[8] That love is real to us, but there is an evolutionary reason why it exists. We also co-operate with and learn from other humans who are members of our tribe, whether it be in our local community, sports team, school, workplace, profession, age group, religion or nation, in order that we might survive, prosper and, ultimately, pass on our genes; and this evolved and cultural tribalism is retained even in individuals who opt not to reproduce. We also fight social, economic and physical wars with other members of the human species to protect what each of us perceives to be the collective good. Other humans are critically important to every one of us, hence it is a simple consequence of evolution that you and I respond more strongly to humans than to other animals and plants; we are predisposed to treat humans as separate from the rest of nature. Members of our own species represent our potential mates, offspring, collaborators and enemies. But this sense of being separate and special is not unique to humans. African lions, American bison and killer whales respond strongly to other members of their own species for exactly the same reasons. Members of their own species represent potential mates, offspring, collaborators and enemies. Every species is special to itself because the survival of each individual's genes depend on it.

This presents us with intellectual conflicts between our rational, instinctive and cultural understandings of the world. Our rational assessment concludes that the natural principles of physics and chemistry generated an increasingly complex set of 'evolved', self-perpetuating chemical reactions, which we call biological life, with no ultimate purpose or intended future. Humans are part of that. Our instinctive or evolved logic, which is then usually reinforced by our culture, is geared to ensure that we propagate our own chemistry by keeping safe, defending ourselves, reproducing successfully, protecting our offspring and relatives, assisting those who might in turn assist us and bartering with those who have something of value to us. Most of the time, this internal or instinctive logic dominates our thinking and behaviour because it is what brought about our existence. No surprise, therefore, that most present-day humans still distinguish between humans and the rest of nature.

Our evolutionary predisposition towards ourselves and other people makes it far easier for us to develop philosophies in which

humans are (somewhat) separate from nature than it is for us to recognize the truth: that we evolved completely naturally, we are still
animals, and everything that we do to the rest of the world is natural.
We may not be happy about some of the changes that are taking place
as a consequence of our existence, but they are still natural.

Not only did we evolve naturally, but it is self-evident that the laws of
physics, chemistry and biology were not revoked when humans
turned up – we are simply using them to our own ends. Think of
some of the changes which might lead people to conclude that our
impacts are unnatural. Yes, we drove many of the largest land animals to extinction, but this is not new. Many large animals became
extinct when North and South America came into contact with one
another, long before humans were around. In fact, mass extinction
events over the last half-billion years typically extinguished the largest species. These human-caused extinctions are simply a consequence
of us acting as an ecological predator.

Next comes transport. Humans have accelerated the rate at which
the seeds of plants and animals are moved around the world in recent
times, but we did not invent long-distance travel. Building aeroplanes
is completely novel, but flying is not; we are just the first animal to
develop a physical tool to do so (unless one counts small caterpillars
and spiders that suspend themselves from threads of silk and are
transported by the wind). Flying birds moved seeds, mites and diseases 50 million years ago, long before we were a twinkle in the eye
of our proto-ape ancestors. Similarly, ships and vehicles have hugely
increased the rates at which species are moving between different
parts of the world, but we did not invent swimming; remember that
tortoises made it to the Galapagos Islands, presumably by hanging on
to floating vegetation. On land, the movement of plant seeds was
once mediated by herds of elephants but is now achieved by humans
and their vehicles, and by the horticultural trade. Why is transport by
one mammal (elephant) more natural than transport by another
(human)? Humans are simply acting as dispersal agents for other animals and plants – a completely natural process.

The consequence of humans changing the climate is that species
are gradually shifting their geographic ranges towards the poles and

to higher altitudes. Again, this is not new. On each of the many pre-vious occasions when the climate has changed (for both physical and biological reasons), the world's species moved. Lastly, we come to transforming the land. Dinosaurs were quite effective at transform-ing habitats before giant rhinoceros and then elephants came on the scene, only to be deposed by humans, who are now the main agents of disturbance. When humans harvest fields of wheat and corn, it is not fundamentally different from the activities of leaf-cutter ants. These amazing insects collect leaves and grow fungal crops in under-ground nests, then harvest the nutrient-rich growths of the fungi to feed their developing grubs. When we drink cow's milk, it is compar-able to other ants that drink the nutritious and sugary exudates of blackfly herds, which they tend like miniature shepherds. When we build huts, houses and dams, it is no different to beavers doing the same. When we use new genetic technologies to move genes from one species into another, it is an extension of the transfer of genes that has previously been accomplished by hybridization, and by microbial and viral parasites. We have just taken these same things to new levels. Of course, many of the specific materials we use are new, and the com-plexity of our tools and communication systems are without parallel, but we have not invented entirely new biological processes. Even if we achieve this in the future, it will still be a consequence of our prior natural evolution.

The upshot of all these quantitative (but not qualitative) changes is a new natural world order in which all manner of populations and species are living in locations they did not previously inhabit. Novel biological communities have come into existence, from the dust mites that live in our beds to the microbes and pelagic invertebrates that attach themselves to fragments of plastic, which float across the world's oceans. An enormous diversity of plants, invertebrates and even vertebrates are living and evolving in cities,[9] and we have planted entirely new forests for wood. European plants are now growing across vast swathes of North and South America, thanks to us mov-ing them there, and they cannot be repatriated. Moths deep in the protected jungle of Sabah's sacred Mount Kinabalu have moved upwards as humans have warmed the climate, and remote reefs are changing where they face higher temperatures and the increasing acidity of the sea. Forest trees growing in the depths of the Amazon jungle have experienced the loss of the largest mammals that used to eat them, growth-altering increases in the levels of carbon dioxide in the atmosphere and physiologically vital changes to the climate. The footprint of human impact is ubiquitous, yet the natural processes of ecology and evolution are still operating in every case. From a biological perspective, individuals are still born and die, populations grow and decline, and evolution takes place. These are regular, natural processes. Why would we regard these new, human-altered ecosystems as any less natural than the ecological and evolutionary processes that are still operating within them?

The world is one in which the specific combination of species and genes in any one place is new but the fundamental biological processes that are in operation are the same as before. While I am describing the current state of the world in this sentence, I emphasize it because I could describe any past period of environmental change with identical words. It would be equally apt for each of the twenty or so great swings in the world climate that have taken place during the last million years. In other words, we can describe human and non-human impacts in exactly the same terms. Species have always moved and evolved, and they have done so particularly rapidly whenever the environment has changed, whatever the cause.

Accepting that ecological and evolutionary change is how nature works means that we must contemplate life as a never-ending sequence of events, not as a single fixed image of how it looks today. This dynamic perspective of life on Earth allows us to put aside most of our doom-laden rhetoric and recognize that the changes we see around us, including those that have been directly or indirectly engineered by people, are not necessarily fundamentally better or worse than the ones that went before. They are just different. We can enjoy and make use of species wherever they might now live, appreciate new Anthropocene species that are coming into existence, and only try to fix things that we, as humans, really think need fixing. We do not need to fix things simply because they are different.

In any event, there is no point crying over spilt milk. The human and non-human contributions to every ecosystem are already inseparable. The practical reality is that the only places where we can attempt to protect the world's 'natural' species from further losses are in bits of the world that have already been altered by humans. Even if we were to remove every human from the surface of the planet today, it would not revert to what it once was.

Standing on the flat roof atop a sixteen-storey former apartment block, the tingling fear of heights set in. I approached the unguarded edge, a high vantage point over the former Ukrainian town of Pripyat, gazing over a post-human world. Swifts, descendants of feathered dinosaurs, wheeled through the air, nesting in crevices in the abandoned buildings. The tops of hotels, schools, a hospital and a Ferris wheel peeked out above the forested landscape, remnants of a utopian Soviet world. Beyond, a brand-new, gleaming steel-and-concrete sarcophagus sat poised, ready to seal in the radiation from Chernobyl Nuclear Reactor Number 4. Like latter-day Mayan temples that became entombed by forest when European diseases initiated the collapse of an extensive former civilization in Central America, Pripyat is a lesson in how ephemeral the impact of humanity can be.

Thirty years after the terrible nuclear disaster that befell Chernobyl on 26 April 1986, Pripyat is a 'lost city' in the making. It is a land from which people have been removed. Visiting troops of tourists appear from time to time to gawp at the shells of buildings from which the valuables have been stolen, and to take emotive photographs of children's possessions that have been callously arranged for this purpose. Then there are security personnel and thousands of workers who still toil in radiation-limited shifts to decommission and encase the former nuclear reactors. Apart from that, a handful of elderly peasant farmers have returned to smallholdings to live out the remainder of their lives. It is not completely unpopulated, but it is close. There are also some radiation hotspots, particularly just downwind of the original disaster, but these quite localized areas are embedded in a far more extensive landscape: 2,600 square kilometres on the Ukrainian side of the border, and a further 2,165 square kilometres in Belarus. Geometry dictates that well over 90 per cent of the wildlife lives in places where radiation levels are tolerable and from where humans have moved out. This wider landscape is a window on to what would happen to the Earth if humans were to depart.

As I peered from the dizzying summit of the apartment cliff, the Trinidad-sized exclusion zone was punctuated by monumental relics of humanity rising above a verdant sea of green. Just a few kilometres away from the failed reactor, falcons rear their chicks on ledges in a derelict cooling tower, their screeching calls echoing in the great, empty chamber. Enormous, slippery catfish gobble up the offerings tourists drop into the canal that was dug to pipe water from the cooling systems. Fishing prohibited, they can reach their full potential. Red-backed shrikes, with their hooked beak, rusty-coloured shoulders, grey head and black mask, fly down to catch grasshoppers, lizards and small rodents, and impale their prey on thorny shrubs. Successfully immobilized, these butcher birds then shred their victims into digestible lumps. Iridescent emperor butterflies jostle with gold-and-brown hornets for the sweet exudates from an oak tree which marks the entrance to an abandoned children's nursery that has sunk beneath the emerald canopy of the entombing forest. The landscape

The former Ukrainian town of Pripyat, with the Chernobyl nuclear reactor and its new cover in the background. Nature is taking back the land, but the forest that is returning includes North American as well as European trees, and Asian raccoon dogs seek meals on the abandoned streets. When humans leave part of the world, it does not return to a pre-human state.

has become home to substantial numbers of wild boar, moose, deer, bear and wolves,[10] although they lie low during the heat of the day. Wild – albeit hybrid – European bison have been released. A black stork wheels overhead.

Thirty years after the event, the benefits of removing human activity from most parts of the landscape can be seen. To a casual observer, nature is returning to how it 'should be'.

Yet it is not reverting to a pre-human world. Erect horticultural cultivars of urban poplars now form dense stands around Pripyat. They push up through the former pavement and their saplings emerge from cracks in the concrete, half a dozen storeys from the ground. They are joined by invasive box elder, a kind of North American maple, that contributes to the regrowth. The inhabitants of the Soviet empire welcomed horticultural offerings from their foes, it seems. The suckers of American locust trees form thorny groves in former municipal gardens and render sports fields invisible. The reversion to forest is being hastened by an array of foreign as well as European plants. Raccoon dogs from eastern Asia prowl beneath these North American trees. In the absence of the extinct European tarpan, conservationists have released endangered Przewalski's wild horses into this weird landscape, despite the fact that these animals come from the grasslands and semi-deserts of central Asia. Extinct forest elephants and giant beavers that would once have helped to maintain openings in the forest, which would have suited the horses, have not existed for a long time. They are not available to return.

Take humans away, and what do we see? The atmosphere and climate have been altered and cannot be buffered from the rest of the world. The land is not reverting to a pre-human version of pristine. Introduced animals and plants from distant continents are here to stay, and will evolve into European versions of their American and Asian relatives.[11] These are permanent biological gains, while large mammals that we extinguished in the distant past are permanent losses. This is unnatural, some might say, yet all the species that are thriving in this landscape are perfectly natural species. The practical reality is that the future of the Earth has already been permanently altered because humans have existed. We cannot unpick it all now.

*

We have been changing the planet for too long to go back. The escalating impacts of *Homo* have been taking place for some 2 million years in Africa[12] and a million in Asia. Around 46,000 years ago, Australian aboriginals eradicated two-tonne *Diprotodon* marsupials and their kin, as well as half-tonne birds and seven-metre reptiles,[13] heralding a global slaughter. Cave paintings represent our encounters with giant beasts, some now extinct, some alive. The first known animal painting is a pig, whose 35,000-year-old image survives in a cave at Maros in Sulawesi. Wild pigs still survive in Sulawesi, but the 30,000-year-old Ardèche hyenas represented at the Pont d'Arc cave in southern France were not so lucky. Our ancestors who sheltered in that limestone gorge must have wiped them out – or extinguished the animals they preyed upon. Today, we think of hyenas as exclusively African, but this used not to be the case. European rhinoceros went the same way, and they are no longer around to repopulate Chernobyl. Waves of human invasion continued to exterminate large animals across Asia, and then in North and South America. By ten thousand years ago, we had killed off most of the world's largest land animals, long before we built our first city, or started to write.

The timing of events meant that the largest animals were near enough gone, or at least on their way out, by the time the climate warmed at the end of the last ice age. When the climate did warm, the world's forests, grasslands, deserts and tundras migrated across the planet's surface, but without these megabeasts in tow. Modern-day elephants can transform the vegetation, and extinct animals would have done the same. They would have altered the structure of the world's forests, savannas, grasslands, marshes and tundras, and hence influenced the types of plants and other animals that thrived in them. The new vegetation that developed after the last ice age is not the same as it would have been in a human-free world. Nor is it the same as any vegetation that went before. Virtually all ecosystems on the land surface have been fundamentally altered by humans for over ten thousand years.[14] The great acceleration of human innovation and impacts in the last few centuries has, in reality, been transforming a world that our ancestors had transformed many millennia earlier. For the two groups of animals for which the best – albeit still crude – historical reconstructions are available, namely mammals and birds, at least twice as many species died out before 1700 as have disappeared since. We should not imagine that recent events are taking place in a previously pristine world. The world has been transformed so thoroughly it is no longer feasible to identify the parts of ecosystems that have been uniquely altered by humans.

The ubiquity of human-influenced change[15] for such a long period means that biological gains associated with our existence are as universal as losses. Yet the gain side of the world's loss–gain equation is often seen as a problem, rather than as a triumph of nature to adjust to altered conditions. Many ecologists and environmentalists, and particularly a special cadre of 'invasive species biologists', are prone to regard changes to the locations where species live as evidence that we are moving towards a less desirable world.[16] They regret how the world is turning out. It is as if there is an 'ought to be' state of the world, with each species having its own 'correct' location. Except that 'ought to' never existed. Nature just happens, and the distributions of species change – no slice of time has any more or less merit than any other. Like it or not, these biological gains will not go away, and more changes will take place in the future. Regarding these changes as unnatural and undesirable is a myopic view of the world.

We need to encourage, not resist, dynamism if we are concerned about enabling nature to accommodate to the human world. That is how our planet's species have survived past change. Too often, we act as if nature is an old master, a great painting that must be kept just as it is. When we perceive nature to be blemished, we attempt to 'restore' it to some past state, just as we might try to repair a damaged masterpiece. To do so requires us to weed out those plants and animals that we think are in the wrong place. We kill successful species to protect unsuccessful ones. It is sometimes possible to do this, but the larger the area we consider and the longer the time period we encompass in our thinking, the more certain it is that we will eventually fail to keep things as they are. But nature itself will not lose. We will simply lose our human fight to return the Earth to one particular romanticized vision of what it might once have been. Like so many Canutes, we have taken on an unrealistic challenge.[17] The tide did not stop for King Canute, and biological change is not about to stop for us.

We can think of the Anthropocene epoch as a fresh start for life on Earth and not only as a passing of the old guard. This is liberating. 'No change' is not an option when we contemplate the future: our choices are all about the direction and speed of future change. We can look forward to future changes with an element of excitement and interest, not just with foreboding. This does not let us off the hook, however. It is entirely within our capacity to turn the Earth into a place that is far worse for humans and also far worse for most (but not all) other forms of life. We need to be vigilant. But simply regretting that things are no longer as they were and venting our frustration at the unnatural state of the world is not the way forward. Let's make the best choices that are possible, accepting that humans are part of the new natural world order.

11

Noah's Earth

Who is to notice the stench of sea lions when enjoying the culinary delights of Fisherman's Wharf: slurping warm chowder, swallowing abalone and savouring seafood stew, accompanied by a bottle of Napa Valley's best? It was a warm end to a chilly day exploring 1073 Lighthouse Avenue, the most peculiar of places, a small park perched above the town of Monterey. Hemmed in by million-dollar homes and retirement pads for the inhabitants of San Francisco, the park contains a remnant of Monterey pine forest. Yet some of these pine trees looked distinctly odd, apparently bearing orange leaves. A new art installation, perhaps? It is California, after all. The billowing crowns of the pines mixed with the long leaves of Australia's gracious gift to the world, the *Eucalyptus*, or gum tree, which also appeared to be covered in orange leaves. Then, as the morning sun rose over the horizon, the leaves opened, revealing the brilliant orange and black of a thousand or more monarchs,[1] America's most iconic butterfly, a few fluttering to the ground to form a golden carpet.

Visitors beware: '*Molesting a butterfly in any way*' can result in a $1,000 fine, according to a 1939 city ordinance. Take care where you tread, lest stepping on one constitutes molestation. Famous for spending the winter clinging to branches in the fir forests of central Mexico in their millions, monarchs to the west of the Rockies mainly head for the coastline of California instead. Monarch Grove Sanctuary, at 1073 Lighthouse Avenue, is one such place. There the butterflies sit out the winter in the not too hot (because they will burn up their fat supplies unnecessarily during the winter if they get too hot), not too cold (so as to avoid being damaged by freezing weather), and not too dry (to avoid becoming desiccated) conditions that are provided by these groves of trees. The oceanic currents that speed past Monterey Bay alternately cool them when it is too hot inland and warm them when it is freezing inland. Having completed their winter snooze beside the sea, the butterflies will then head back up the Cascades and across the Sierras and Great Basin, seeking milkweeds for their caterpillars to eat during the warm summer months, before their great-great-great-offspring (or however many generations it is) return to spend the following winter back in California.

This mild climate, which attracts both monarchs and humans to the Californian coast, is also ideal for Monterey pines, although the pines are even fussier in their requirements than the butterflies. They do need some moisture, but it has to be dry enough for occasional fires to spread through the forest canopy. Fires can kill the mature trees but at the same time they release seeds from the tree's resin-sealed cones to start the next generation. Without heat, the seeds are entombed, and there would be no new seedlings. Fire is therefore an essential part of the ecological system that allows Monterey pines to survive from one generation to the next. The local homeowners in suburban Monterey are less than enthusiastic about reducing their properties to piles of ash, so converting the forest into a bonfire meets with a certain level of resistance. The seeds require an alternative means of escape. This is where the local authorities have stepped in: rather than torch the neighbourhood, they collect the cones, extract the seeds and deliberately plant them to establish the next generation of forest trees.

While occasional fires are essential, the Monterey pine does not like it to be too dry because then the fires become too frequent. If the trees burn to the ground while they are still too young to produce new seed capsules, the generations will be interrupted. The trees disappear without replacement. Their not too hot, not too cold (and not too wet, not too dry) lifestyle makes the West Coast's Goldilocks tree one of America's rarest. It survives on a handful of coastal Californian hills, and on the Mexican islands of Guadalupe and Cedros, off the coast of Baja California. Already threatened by urbanization, fire suppression and pine pitch canker, a fungal disease that has been imported from the south-eastern United States, human-caused climatic changes could now push this tree over the edge. The twenty-second century could mark the Monterey pine's final hurrah.

Monarch butter-
flies overwintering
in California: rest-
ing and sunning
themselves while
sitting on leaves of
Australian-origin
Eucalyptus *(above),*
and flying among
shiny-barked Euca-
lyptus *trees (right).*

The Goldilocks tree has been sensitive to the climate throughout the last million years of ice ages and intermittent warmer periods. Over this period, pollen blown out to sea and material washed down California's Santa Clara River has sunk to the bottom of the Santa Barbara Basin, where it has become incarcerated in layers of marine deposits. There it waited patiently until geologists and ecologists hired a boat and set sail across the basin, drilled holes into the deposits, pulled up the cores, then counted the pollen grains in each layer, allowing scientists to track changes to the tree's abundance as the region's temperatures waxed and waned. In the words of Connie Millar of the US Forest Service: the Monterey pine *'was least abundant during full interglacials [relatively warm periods like today], when oaks dominated coastal habitats, and was also uncommon during the cold periods of the glacials [full ice-age conditions], when junipers dominated'.*[2] The problem today is that we are already in a too warm interglacial period, when the trees are at their rarest, and it is getting warmer still, which means that the climate is becoming even less suitable for the Monterey pine. Under attack from concrete, canker and now climate change, the Goldilocks tree could disappear completely.

Fortunately, some inspired – or perhaps just lucky – foresters have been on the case. The Monterey pine, whose scientific name is *Pinus radiata*, was included in a trial of tree species that was carried out in Canterbury, on the South Island of New Zealand, in the 1850s.[3] It seemed as though the tree had been yearning for antipodean weather all along. New Zealand's moderate climate, bathed by the waters of the south Pacific, was ideal for the sensitive tree. Growing about seven times faster than back home in California, radiata, as it is now known, has become the mainstay of New Zealand forestry, covering some 18,000 square kilometres of plantations. This amounts to nearly 90 per cent of New Zealand's forestry and represents about 4 per cent of the entire country's economy.

And it didn't stop there. Radiata now generates around 95 per cent of Chile's timber production and forms the largest part of Australia's plantation wood. It is grown in Argentina, Kenya, Uruguay and South Africa, where the climate is so ideal that the tree has gone wild, colonizing grasslands, parts of the native fynbos vegetation and forest clearings in the mountains. Thanks to foresters, the Goldilocks tree has discovered that there are plenty of other places on planet Earth where it can grow. A species perched on a few coastal cliffs in California has become a global colonist – an endangered species converted into an heir to the world. In Australia and New Zealand, it has even been joined by the monarch butterflies, which have been introduced there, too, along with the milkweeds their caterpillars eat. Although there are justifiable fears about the survival of some of the overwintering sites for North American monarchs, the butterfly itself is one of the most successful in the world. It lives in the Americas, and now, thanks to us, it has become widely established in Southeast Asia, on many Pacific, Atlantic and Indian Ocean islands, and in southern Europe, as well as in Australasia.

Meanwhile, back in Monterey, blue gum trees, *Eucalyptus globulus*, have made the journey in reverse. Originally confined to the island of Tasmania and to the southernmost parts of Australia's Victoria state, they pose a particularly tricky Anthropocene conservation challenge for those who live around Monarch Grove. According to the California Invasive Plant Council, they are actively regenerating along the coast, represent a fire danger and oust native plants. Are they foreigners that should be evicted, or attractive trees that are well suited to California's coastal climate? The paradox is heightened because the larger leaves of the blue gums – compared to pine needles – form just the right conditions for the Monarch butterflies to settle down for the winter.[4] Planting more gum trees may be one of the best ways to help protect North America's most iconic butterfly.

This is the antithesis of traditional conservation thinking. Australia's blue gums are thriving in California, Hawaii, Ethiopia, southern Africa, Morocco, Spain, Portugal, Cyprus and elsewhere in southern Europe, right across to the Caucasus in Georgia. Giant wombats, Tasmanian wolves and disease-sensitive frogs that evolved in isolation in Australia have generally succumbed when they have faced up to the virulent inhabitants of Europe and Africa. Not so the blue gum, which today thrives on all the world's major landmasses, save Antarctica.

Monterey pine and blue gum were both confined to small areas of the world's land surface until humans evolved and transported them

Monterey pine, Pinus radiata, *clings to cliff edges to the south of Monterey, in California, where it is a rare and potentially endangered native species of tree (above). It also grows as serried ranks in plantations in New Zealand (below). Radiata, as it has become known, has been plucked from obscurity to become one of the world's most important commercial trees, with a wide variety of uses, from pulp for paper-making to timber for house-building.*

elsewhere. They illustrate the current importance of some formerly rare species to humans, and we might presume that some presently rare species will be valuable to us in future – although we do not yet know which. Both trees could have been threatened by human-wrought changes to the vegetation and climate.[5] And yet they have become widespread, to such an extent that there is no risk that either of these species will be extinguished for the foreseeable future. Gum trees seem destined not only to survive the human era but potentially to become a globally important (rather than just Australasian) type of tree for tens, if not hundreds, of millions of years to come. Forests of the new natural world will never be the same again. Environmentalists may dislike them for their new-found success,[6] but there is no doubting they represent previously rare species made good.

Rare species becoming common and common species rare is not a human-created phenomenon that has been invented in the last few centuries. It has happened repeatedly over the last 2 million years as the Earth's climate has zigzagged back and forth between glacial and warmer climes. The alpine chough is a perky yellow-billed and glistening-black relative of the crows that today wheels in a dancing flight over meadows at high altitudes in the European Alps and Pyrenees, but it used to be a more frequent denizen of the lowlands during the depths of the last ice age. We know this because our cave-dwelling ancestors, who hunted over treeless ice-age landscapes, did not put their rubbish out, leaving chough bones to be discovered by archaeologists who carefully sifted through their twenty-thousand-year-old trash.[7] And this is likely to be true of many other mountain plants and animals that are adapted to colder conditions, and which were much commoner during the ice ages than they are today – as was the case with the British dung beetles that Russell Coope studied, one species of which now survives only in the Pyrenees, and another in Tibet.

While alpine choughs and many others were thriving under the colder and drier climate of twenty thousand years ago, oak, lime, beech and hornbeam trees were banished from the European prairie – the frigid, open steppes of Asia extended right into Europe at that time. These trees survived in sheltered valleys and on sunny slopes, mainly in the Balkan states, and in the Italian and Iberian peninsulas of southern Europe. They were quite localized plants, just as the Monterey pine was rare until humans came along and redistributed it across the southern hemisphere. Yet these European forest trees survived and, when the Earth heated up, a little over eleven thousand years ago,[8] they spread out and colonized the land. Once-rare trees sprang up across the continent and became the dominant vegetation: the foundations for the new forest that grew around Lake Maggiore, spread to France and the British Isles in the west, reached Chernobyl and even further east, and gained a footing in Scandinavia in the north. The summer-green forests of oaks and limes and hornbeams that would clothe most of Europe (if humans had not cut most of them down again) did not come into existence until at least thirty thousand years after modern humans colonized Europe.

Similar stories can be told for most of the other extensive forests of the northern hemisphere. The conifers that form Canada's great boreal forest, for example, moved in only once the ice retreated at the end of the last great glacial period. When foresters and conservationists refer to primeval forests that were inhabited by European bison and aurochs and boar, and by woodpeckers and dormice, they are talking about places that used to be grassy steppes where our ancestors hunted. These forests are primeval only in the sense that the cultural knowledge of our steppe-dwelling ancestors has been lost, making them seem ancient and enduring to recent generations of humans. Those ancestors may well have been rather irritated by the arrival of trees. The trees inherited this warmer land from the grassland and (now) mountain plants that went before.

It follows that everything that we, as humans, obtain from those forests has also arrived recently: the beasts of the chase, the fruits we harvest and the forest-dwelling fungi. These trees gave us somewhere to live by providing timber for the construction of our houses and furniture, food by developing the continent's soils, which we subsequently turned into cultivated fields, means of storing our food in wooden granaries and barns, and wooden bowls to eat it off, and materials to corral our domestic livestock in wooden stables and behind fences. They provided our transport system and means of cultivation: wheels, carts, ploughs and ships. And they gave us our heating and ability to cook. Less gloriously, they provided the materials of war: wooden spikes and fortresses to keep out the enemy, the shafts of spears and handles of swords, and the bows of archers.

Converted to charcoal, the trees not only allowed our ancestors to cook food, but enabled Stone Age people ('Wood Age' might be more accurate) to smelt the first copper seven thousand or more years ago. A thousand years later, charcoal was used to heat alloys of copper, tin and arsenic, ushering in the Bronze Age, followed by iron and steel smelting in the last three to four thousand years. The Iron Age had arrived, thanks to charcoal, thanks to wood, thanks to the trees whose rare predecessors sheltered on the warm slopes of the Balkan forest. We were comfortable: cosy in front of a warm hearth in a wooden house, sitting on wooden furniture draped in the fur of forest animals, eating cooked meats and baked bread from a pottery bowl fired in a charcoal-heated kiln. If we were rich enough, we drank out of a charcoal-smelted glass beaker. Rare trees became common and then gave us our way of life.

The forest limes, beech, hornbeams and oaks of Europe, and the radiata pine and blue gum trees that grow above Monterey Bay, all show how species that are rare at one time can become common at others. This has always been the way of nature. The majority of species that have ever become really important contributors to the Earth's ecosystems have evolved in a relatively localized area first, before they spread more widely, and then their numbers have waxed and waned as conditions have changed. These examples also illustrate that rare species can go on to be of immense economic and social benefit to humans. Most of our carbohydrates come from plants that are descended from a few species of grasses that were restricted to small parts of the world before people arrived: maize, rice, the ancestors of wheat, and so on. Cattle and pigs were not particularly rare, but nowhere near as common or as widespread as today, and sheep and goats and chickens were quite ordinary species that lived in different parts of southern Asia. They now provide most of the meat protein that nourishes the world's human population, and they can be found nearly everywhere that people live.

The list goes on when one considers all the good things that we have obtained from other unassuming plants (medicines, insecticides, carbohydrates), microbes (brewer's yeasts, antibiotics, methanogens in bioenergy plants), and animals (honey, pets, meat, leather). Rare species made good is not just a feature of species that we actively use. Like the Bactrian house sparrows and house mice that originated in Asia, and humans in Africa, many other animal species started off living in a small part of the world but have since taken the Earth by storm. Similarly, European herbs that used to live in a few small mountain ranges or ravines now grow widely across parts of North America, after people transported them there.[9] Mount Etna *Senecio* – transformed into Oxford ragwort – now populates every British town.

Huge changes in the commonness and rarity of different species are a feature of the entire history of life. Now, we are seeing changes again as humans alter the world. Some rare species have become common, while lists of the world's endangered species highlight those that are increasingly rare and may be lost entirely.

The inevitable ups and downs of different species that take place when the environment changes, and as they colonize new locations and evolve new characteristics, dictate that we will ultimately fail if we attempt to keep things exactly, or even roughly, as they are. This dynamic perspective of biological change might sound like capitulation, but, in fact, it releases us. The Earth was not in some perfect or final state before humans pitched up. Life is a process, not a final product. So we need a conservation philosophy that is based on natural change, with humans centre stage: partly because we have already brought about so many changes to the world that cannot be ignored, and partly because humans evolved naturally and we are part of the natural system.

Such a philosophy has four overarching principles.

The first principle is to accept change. Deviations from the past state are not all 'worse'. We must recognize that the diversity of life on Earth is determined by the balance between gains and losses, and it is just as legitimate to maximize gains as it is to minimize the rate of loss.

The second principle is to maintain flexibility for future change. There is no single way to achieve this, but, although it might seem paradoxical, the most important contribution we can make is to save the world's existing species – within reason (a topic that I will return to shortly). We can think of them as currently rare species that may in future become common – the Earth's spare parts that might be needed in the future, when new events unleash the next stages of environmental change.

The third principle is that humans are natural within the Earth system, so anything we do is also a natural part of the evolutionary history of life. We can be adventurous and use whatever technological or other strategies might be available to us to ensure that we hand an operational Earth on to future generations, without fear that we will somehow make the world less natural.

And the fourth principle is that we still have to live within our planetary bounds. We know that we cannot expect the bounty to continue if we carry on killing animals faster than they can breed or cut forests down faster than they grow. This strategy failed when our ancestors drove most of the world's largest land animals to extinction, and it has played out in the last few centuries as whale and fish populations have collapsed under the pressure of over-harvesting.[10] We need a resilient and sustainable approach. We should aim for maximum efficiency, by which I mean that we should pursue strategies that fulfil all human needs – and, where possible, desires – of every citizen on Earth while generating the least possible collateral damage to the global environment.

Converting this philosophical quadrumvirate – accept and promulgate 'good' change; maintain flexibility; use any means available because human actions are natural; live within our planet's bounds – into agreed practical strategies and actions is far more difficult, not least because individual people and societies will differ in their preferences. I will just touch on a few of the options.

The first principle is to accept change because this is the natural way that the biological world responds when the environment changes. Accepting change is not the same as a laissez-faire approach. It is instead about considering the pros and cons of possible alternatives, knowing that 'no change' is not an option. It is about prodding the world in a desired direction as effectively and efficiently as possible.

Being efficient usually means that it is a far greater priority to address the underlying causes of environmental change, such as generating energy without causing climate change, or obtaining sources of protein in the human diet from the minimum possible area of land and sea, than to patch the world up afterwards. For example, once we are able to produce laboratory-grown animal muscle economically and generate 'meat substitutes' from plant and fungal material that people *really* like as much as meat, killing living animals to derive our sustenance might become socially unacceptable. This could potentially be achieved rather soon, by which I mean within fifty years to a few centuries. At that point, the pressure to convert more land into pastures would be greatly reduced.

If the cause cannot be rectified, there needs to be an extremely good reason to embark on an indefinite treatment of symptoms. We need to be able to justify keeping unstable ecosystems in their current state or saving species that are no longer viable. We might decide to put in the effort, as New Zealand currently is, to protect pedestrian birds and crawling bats on account of their cultural and scientific interest,[11] but if we decide that they fail the ecological and evolutionary triage of modern life, the future trajectory of our biological world will not be altered in any important way. Our aim should be to maintain robust ecosystems (however different from those that exist now or existed in the past) and species, rather than to defend an unstable equilibrium. We can let change happen.

At present, our default position is to treat change as negative. The nations of the world have agreed that we should aim to save biodiversity, signing up to the overarching international framework of the Convention on Biological Diversity. The representatives of each nation go back home from their international congresses, charged with at least slowing the rate at which biodiversity is lost within their own country. This requires each country to establish what it already has.[12] Loss has to be measured against something, so 'the present' (or recent past) represents a baseline against which any future change can be measured. That seems sensible. If any 'native' species or 'natural' habitat declines within a country, it triggers concern.

The convention has been immensely beneficial, but the setting of baselines for species and habitats, and the calculation of trends separately for each country, has had a very negative side-effect. It has formalized a no-change-is-best framework for conservation through-out the world, when we know that dynamism is how species ultimately survive periods of environmental change. By saddling our assessment to fixed baselines within national boundaries, all changes, including gains of new species that arrive from other countries, represent devia-tions from the baseline that has been set. A decline of a species in one country registers on our biological accounts as negative, and its arrival in another country is likely to be either ignored or counted as negative (as a biological invader, which is also recognized as a prob-lem within the convention), even if the overall international status of the species is unaltered. This can't make sense.

It is difficult to understand why any particular moment in the con-tinuous passage of time should have special significance. Why are the dates of conservation congresses in the late twentieth and early twenty-first centuries superior to other baselines? Why not go back 130,000 years ago for a model of how the human-free world should be? This was the last time the Earth had a similar climate to today, in the period before modern humans had emerged from Africa and killed off the large animals. The fundamental flaw of baselines – or backwards-facing conservation – can be appreciated once one real-izes that every ecological and evolutionary gain that took place prior to a particular baseline date is defined as desirable, whereas all the gains that took place afterwards are likely to be dismissed, disliked or repelled (equally, losses before the baseline are accepted, but those after it are not). Arbitrarily move the baseline to an earlier date, and more of the gains are deemed undesirable. Baseline reasoning makes no logical sense when life on Earth is a dynamic process.

It is enormously helpful to understand and draw inspiration from the past, and I have repeatedly done so in this book, but going back to it is not on the table, as we saw in the abandoned landscape that surrounds the now-defunct Chernobyl nuclear reactor. Many of the biological changes of the human epoch are already permanent. We should be informed by the past, but not circumscribed by it. We can only go forwards, and that means dealing with species moving to new locations, with evolutionary change in those species, and with new species that come into existence.

*

The second principle is to maintain flexibility for future generations, in the knowledge that we cannot predict the future with any accuracy. It is no easier for me to imagine the world that my descendants will inhabit 150 years from now than it was for my great-grandparents to have imagined the transformations that have taken place between the 1860s and today. How can we tell what new politics, social attitudes, construction, transport, pollutants, means of obtaining our food and drink, biomedical products, diseases, artificial life, artificial intelligence, robotics, energy sources, weapons systems and unknown unknowns will represent new global norms by the second half of next century? Nor can we know how we and the biological world will respond to these changes. This is sobering when we consider questions such as whether, when and how we might deliberately intervene to protect nature. While it is perfectly acceptable to make short-term decisions that will bring pleasure, health or other benefits to those of us alive today, an underlying philosophy of conservation is for us to act as custodians of the natural world for future generations.

Thinking beyond the direct and current needs of people who are alive today, any human-oriented conservation strategy must be about maintaining opportunities within the biological world of which we are a part. To be able to take full advantage of the world's biological potential in future, we should keep alive the building blocks of the biological world. By building blocks, I mean those populations and species that will form tomorrow's ecosystems. The more different kinds of species exist, the greater will be the chance that at least some of them will flourish under the new conditions (whatever they are), and it increases the chances that new evolutionary adaptations and hybrid forms will have properties that enable them to contribute to the Earth's biological processes. This is not a preservationist strategy but a resilient, flexible approach that will fuel rather than prevent future biological change. Not only will existing species and their evolutionary descendants form future ecosystems, they will also be the sources of new discoveries. Such is the rate at which new biological technologies are advancing that we are probably entering a period of unprecedented growth in the ways in which we develop and use natural products from living animals, plants, fungi and microbes. If we care about the longer-term condition of humans on Earth, we should not ignore species that are unfamiliar or rare today. These may be species on which humans rely only a few generations hence, just as European civilizations relied on once rare Balkan trees and we now utilize previously rare Monterey pines and gum trees.

Keeping as many species as possible alive is a momentous challenge, particularly when we realize that their future survival will not necessarily depend on them continuing to live in the places where they currently thrive. As the climate changes, for example, perhaps a quarter of all species will become climate refugees, surviving only outside their historical distributions, and nearly all species will live in at least some new places. Helping species to reach these new locations – promoting gains at the same time as minimizing global losses – is likely to become a major focus of conservation in the second half of this century. Keeping them alive *somewhere* on our planetary Ark is the challenge we face. This keeps our biological options open, even if we do not know precisely how each species will contribute to future ecosystems, or whether they will improve the wellbeing of future human generations.

The third principle is that humans are natural within the Earth system, and so it follows that anything we do, or do not do, is a perfectly natural consequence of the evolution of a bipedal ape. We can intervene in ways that old-thinking would define as 'unnatural'. We can be proactive rather than bowed over with regret that things are no longer as they were.

The ginger-guzzling elephants I met in the under-storey of Borneo's Danum Valley forest are a case in point. Imported from a now extinct population that most likely lived in Java, the island of Borneo now holds the world population of this genetically distinct form of Indian elephant. No one, as far as I know, is proposing to remove them – but if they could be proven to be 'native' to the island of Borneo after all, conservationists would then regard their presence as unambiguously beneficial.

If we step back, does the answer to the question of whether they are 'native' really matter? It is difficult to see these elephants as bringing about 'bad change', provided we accept that humans are part of the new nature. Any unique genes[13] that are held by these animals survive only because they were transported from one island, where elephants died out, to another, where they have succeeded. Genes, like species, survive because they keep track of the changing world. With humans as part of the natural system (human-mediated dispersal is natural), the presence of elephants on Borneo now is natural, irrespective of however short or long their history in the region may be. There is no suggestion that they are driving Bornean species to extinction, and their presence may be beneficial by providing dung for beetles and generating small-scale disturbances for plants to regenerate, as well as water-filled footprints where damselflies can breed. It is true that a forest with elephants is going to be different from a forest without elephants, but the net balance of gains and losses is not necessarily going to come down on the side of debit.

Such accidents pepper the world. All the many thousands of transported plants and animals that have established populations in new regions have demonstrated time and again that species may flourish outside their historical ranges. In doing so, they normally increase the total number of species that live in each region, they start to evolve into distinct forms and they sometimes hybridize and create new species. Eventually, they will increase the diversity of life on Earth. Unless one synonymizes change with loss, these events are not inherently bad, so why not use similar approaches in conservation? Moving threatened species to places where they will prosper could save them from extinction, as could the infusion of new genes from other populations or species. The critically endangered yellow-crested cockatoo, for example, is thriving as an introduced population in Hong Kong while continuing to decline in its Indonesian homeland.[14] There are myriad opportunities, but the conservation world rarely countenances such thinking. There is a fear that projects of this type will cause yet more change, but change is going to happen anyway, whatever we do. If we are going to have new biological communities in any case, why should they not contain a smattering of rarities, rather than just the globe-trotting generalist species that will turn up on their own?

One thinker in this area is Josh Donlan, from an organization known as Advanced Conservation Strategies, who has come up with a vision for North America and, by extension, the world. He wants to create big, wild spaces containing big, wild animals. That is conventional enough. The unconventional part is to identify ecologically similar relatives of the bitumen-preserved animals from California's La Brea tar pits and release them in North America. Mastodons, large-headed llamas and American lions may have shuffled off this mortal coil, but we could always replace them by Old World elephants, South American llamas and African lions.[15] Cheetahs could be introduced to pursue native pronghorn antelope in the absence of the original fast-running relatives of pumas that used to populate the plains. Donlan and his collaborators have suggested that having this international cast of animals roaming the wilds would improve the prospects for North American plants, which evolved in the presence of large grazing animals. The flora might be expected to survive better if large animals are returned, with the aspiration that all the native insects that feed on the plants would then prosper, too, and the lizards and birds that feed on insects would thrive in turn.

This is an intriguing idea, but it is perhaps unfortunate that the originators of the idea called the proposal 'Pleistocene rewilding'. The Pleistocene is the geological epoch that covered the last two and a half million years of sporadic ice ages (until 11,700 years ago), so this implies a return to the past. Rewilding (as opposed to wilding) directly conjures up the expectation that we are returning to a more pristine past, to an older baseline, rather than starting off in a new direction. Whatever happens, and however many different species are released, it will not create a facsimile of the past because North America's vegetation has already developed for ten thousand years in the presence of humans and the absence of the extinct animals – and the replacement animals are not exactly the same species as those that became extinct. Vast ground sloths and sabre-toothed *Smilodon* cats have no modern equivalents that are available to release.

Europe is much the same. Twenty-first-century Europe lacks rhinoceros, hippos and elephants because our ancestors killed them all. We could release their nearest living relatives: Mediterranean shrublands could become the refuge of African black rhinoceros; while great one-horned rhinos from India and Nepal and African hippos could find security alongside white horses and black bulls in the wetlands of France's Camargue. Again, great fun, but not what it once was. The past has gone, and who is to say that the future impacts of an international medley of imported creatures in Europe and North America, or anywhere else in the world, would be fundamentally that different from using feral sheep, cows, donkeys and horses? However, it is well worth exploring. The greatest contribution of these ideas is to expand our ambitions beyond conventional conservation, which sometimes feels akin to fiddling while Rome burns. We can think about engineering new ecosystems and biological communities into existence, inspired but not constrained by the past.

Debate about such proposals tends to dwell on three areas of concern. Is it natural? The answer to this question is yes, as I discussed already. Is it practical? There are inevitably great challenges, and the legislative barrier to releasing lions (which will eventually kill some people as well as livestock, even if it is only a few) in North America might be insurmountable. However, it is practical in some locations, at least on private ranches with adequate fencing, although this rather goes against the grain for the originators of the idea, who are dreaming of a magnificent, wilder landscape. Is it better or worse for American biodiversity? This last question is clearly unanswerable. None of the possible future states of the world will be fundamentally better or worse than the world's pre-human condition. They will just be different.

So we need to define 'better' as a human preference. Unfortunately, there is no consensus of what 'better' should be. If you accept my argument that we should minimize the number of species that become extinct because some may turn out to be important in future, then 'better' in the case of releasing rhinoceros species in Europe would not be that European ecosystems are somehow yearning to have rhinos bumbling around in them but that the grazing and browsing they provide might possibly save some Mediterranean plants and insects from extinction. This may be so, but there is no evidence one way or the other – although it could be ascertained by experiment. The more tangible benefit is that rhinos are endangered elsewhere. We could stop them from going extinct, just as Javan elephants have survived because they were transported to Borneo. But if this is the logic, why stop there? If black rhinos are worthy of being saved, why not mammoths?

'De-extinction', or resurrection biology, is a term that is increasingly used by scientists and conservationists to describe exactly this, the latest fad in affirmative action – more brave thinking. If currently rare animals and plants might in future become more abundant and widespread, why should this privilege not be extended to species that have recently become extinct? So far, it has been attempted in earnest only for the extinct Pyrenean ibex, a subspecies of Spanish ibex, albeit without success. Genetic material from Celia, the last living female, was injected into a domestic goat's egg, stimulated to start growing, and then implanted back into a surrogate goat mother. Despite many attempts, however, all the embryos died before birth, except for one kid that had a deformed lung and perished a few minutes after it was born. Even if a Celia II could be created (in principle, it should be possible), there are still no males, so we are a long way from re-creating a viable population. Maybe it would be simpler to let Celia rest in peace and release other kinds of ibex instead – different Spanish ibex subspecies are still available. Taking inspiration from the Bornean elephants and Donlan's ideas, and marrying this up with a strategy to save as many different species on Noah's Earth as possible, a more adventurous strategy would be to release the closely related and highly endangered walia ibex, which is down to a remnant population of about five hundred individuals in the mountains of Ethiopia. That would be a greater contribution to global conservation.

The next 'big prize' in resurrection biology is the woolly mammoth. Several research groups are attempting to bring mammoths back by cloning their permafrost-preserved genes, or by splicing mammoth-like genes (which govern the appearance and particular physiological attributes of mammoths) into Indian elephants to create chimeric elemoths or mamphants. I very much look forward to seeing them; but we should not kid ourselves that mamphants are woolly mammoths. They are hybrids, made in a laboratory – although I do not personally have a problem with this. If we can protect these animals from our traps and guns, and eliminate the trade in ivory (or grow ivory in vats), there is no reason why mamphants could not 'look forward' to a long future of evolutionary success. Knitting the genes of different species together might also work if we wish to breathe life into animals that resemble (but are not the same as) any of the other 177 human-killed large mammals that have died out in the last fifty thousand years but whose ancient DNA is considerably more damaged (not having been frozen) than that of the mammoth. Bison bounced back with a few cow genes on board. Why not animals that are quite like large-headed llamas, if their genes can be dug up from somewhere and spliced into Asian camels?

If this were to happen, and it surely will for at least a few of the extinct species, these spliced-together creatures will represent novel hybrids – biological gains – that may or may not turn out to be successful. They will be similar in this respect to other hybrids that have come into existence in the Anthropocene, such as honeysuckle flies, Italian sparrows and Oxford ragwort, which are successful, or Yorkwort, which is teetering on the brink of extinction a few decades after it originated. To all intents and purposes, we will be creating new hybrid species. This would be an interesting precedent. Although we have repeatedly hybridized domestic animals and, particularly, crop plants for use in agriculture, it would be something of a new step if we are deliberately to create new hybrids and then release them into the wild.

This leads to the next question. Why should we do this only when one of the parent species happens to be extinct? If an elemoth is to be created from one living (Indian elephant) and one extinct (mammoth) ancestor, why not create novel entities from two or more living species (or two or more extinct species)? If the goal is to have a relative of llamas and camels roaming the Great Basin, and for those animals to be unique to North America (as large-headed llamas were), I see no particular reason why inserting a few large-headed llama genes into camels or guanacos is a better option than creating a population of camas, which are hybrids between camels and South American llamas – the domesticated relatives of wild guanacos. Camas would contain the surviving gene pool of the camel relatives that first originated in North America.

The purpose of this discussion is not to advocate or dismiss any one or another of these specific ideas but to recognize that, if we regard all human actions as natural, and thus legitimate options, we can use technologies and contemplate strategies to maintain and increase biological diversity in a variety of ways that ecologists and conservationists would regard as unconventional – and which would historically have been regarded as unacceptable. Aiming to increase biological diversity is just as legitimate as attempting to reduce the rate at which (existing) diversity declines. In the case of mammals in North America, we could adopt some combination of strategies: release ecologically similar species from different continents, release endangered species from elsewhere in the world, import threatened species from Central America as the climate warms, re-create nearly 'pure' forms of a few extinct species, genetically splice ancient DNA and living species together, generate new hybrids by conventional breeding (as in my 'cama' suggestion, and the now vilified beefalo hybrids between cattle and bison), use domesticated animals to graze and browse, or just leave wild deer and feral pigs to get on with it, and allow grizzlies and wolves to recolonize from the north, as they already are.

The conservation rationale for each of these approaches is usually stated to be some combination of making the world more natural (which is fallacious) and restoring it to a state that resembles one that existed in the past (which is equally fallacious). None of these options will cause the biological world to revert to its original trajectory. However, we should not exclude any option simply because it is novel. It is up to us to decide whether we wish to intervene and, if so, how. Any such decisions, including the decision not to act at all, will simply set the world going in a slightly different but equally natural direction.

Back in Monterey, it is again possible to take in the bizarre sight of America's most famous butterfly surviving the winter by clinging to the pendulous twigs and leaves of an Australian gum tree. Beside it grows a rare native Californian tree that is now a mainstay of the southern hemisphere's forestry industry. The Monterey pine is saved, even if its eventual survival may depend on populations that have gone wild in South Africa. No, this is not how nature used to be before humans came along. But we did, and this is the new natural world. Species survive where they can, thrive where they can, and are used by humans where we are able to use them.

California now supports well over a thousand species of introduced plants and, although the ice age giants that lie preserved in the tar pits at La Brea are gone, the last few hundred years has seen the addition of European rabbits, two species of squirrel, house mouse, two rats, Barbary sheep, fallow deer, wild boar, pheasant, turkey, starling, house sparrow, brown-headed cowbird and numerous insects and slugs, as well as all our domestic animals. Like the star-thistles that are already on their way to becoming new species that live only in California, all these introduced plants and animals will begin to diverge from their ancestors which still live in other parts of the world, and eventually increase rather than decrease the variety of life on Earth. It is not necessarily a tragedy for the biological diversity of California, or for our planet as a whole. We should still try to save species that live only in California, but waging war on interlopers is unlikely to be important except in situations where they specifically endanger other species. Even then, it is worth the fight only if a permanent solution is available.

Whenever our urge is to fight a specific biological change, we should ask the following triplet of questions. Will our efforts have made much difference a few hundred years hence? If not, this means we are fighting a battle we will inevitably lose. Next, will our great-grandchildren's great-grandchildren be that bothered if the state of the world has been altered, given that they will not know exactly how it is today? If the answer to this second question is no, this means we are fighting battles we do not need to win. If change is inevitable, which it is, we should then ask a third question: how can we maximize the benefits that our descendants derive from the natural world? In other words, how can we promote changes that might be favourable to the future human condition, as well as avoid the losses of species that might be important in unknown ways in future?

If we can create new biological success stories by whatever means, let's do it. We can protect animals and plants in places where it is feasible to do so, rather than where they came from. We can transport climate-threatened species to places they could not otherwise reach – why not, if this increases the chances that individual species will survive? We can import species into ecosystems where they did not previously occur, for example if drought-resistant trees could increase the resilience of a forest to future water shortages. We can introduce species to new geographic regions so as to increase the impoverished diversity of human-created habitats. We can foster novel ecosystems that contain mixtures of species never seen before. We can deliberately create ecologically diverse landscapes that are mosaics of different kinds of ecosystem, richer in species than most that exist today. We can also help direct the evolutionary process: establish new hybrids that will perform ecological functions we find useful, develop new forms of insects that will eat pestilential weeds, and use genetic modification technologies to insert disease-resistance genes into captive frogs so that they can repopulate South America. It is time for the conservation and environmental movement to shed its self-imposed restraints and fear of change and go on the offensive.

Welcome to Anthropocene Park. Our home.

Epilogue
One million years AD

Surprising as it might seem, it is entirely possible that the long-term consequence of the evolution of *Homo sapiens* will be to increase the number of species on the Earth's land surface.[1] Of course, it is impossible to anticipate what might happen in the future. Humans may become extinct as the result of an event so catastrophic – for example, an extended nuclear winter, or the biological collapse of the planet associated with artificial life – that all bets are off. And who can guess what will unfold if humans or our bio-mechanical upgrades survive the next million years and develop currently unimaginable technologies? However, if we extrapolate from the biological processes that we do know about and ignore the world of unknown unknowns, it is at least possible to contemplate the impact of humanity on life on Earth a million years hence.

Consider the losses first. We have already extinguished most of the largest land mammals, and we have removed most of the flightless and disease-susceptible birds that used to live on isolated islands. Other island species that are unable to survive in the presence of continental species are also on the way out. In future, additional mammals, birds, reptiles, fish and plants may also be hunted, fished or picked to extinction. Some localized parts of the world will lose many of their unique species if they become so heavily populated or cultivated that humans remove almost all vestiges of the former vegetation. Many high-mountain species will disappear as the climate warms. Undoubtedly, there will also be losses that we cannot anticipate.

How much will all this extinction come to? Heading the list so far are the birds, which have already lost around 9 per cent of the species that used to exist, most of which were former inhabitants of oceanic islands. Additional species will have been lost without trace. This is not typical, as far as we know, but it is possible that Pacific land snails and some groups of insects will have suffered as much. However, when it comes to vertebrate extinctions so far, it seems that birds are top of the list (amphibians may catch up). At the current rate of extinction, we can expect another 12 per cent of the world's bird species to disappear in the next thousand years of human existence.[2] The overall average extinction rate (not just for birds) seems to be heading towards the loss of around 10 per cent of species in a thousand years, which falls far below the level of extinction (75 per cent plus) required to match one of the previous 'Big Five' mass extinctions in the geological past. However, the Anthropocene would qualify as the 'Big Sixth' if the current rate of extinction keeps up for the next ten thousand years, which is a very short time in the history of life.

Declaring the Anthropocene to be the sixth mass extinction is somewhat premature. Nonetheless, a quarter of mammals, a third of cacti and reef-forming corals, and two out of every five amphibians are listed as being threatened in some way.[3] They will not *all* die out, but some threats are likely to escalate in the near future, such as human-caused climate change and more intensive use of the land linked to the human population. So, based on recent rates of extinction and projected threats, we could guess that human-linked extinctions will end up in the ball park of 20–40 per cent in the next thousand years – for most kinds of animals and plants. That might be about it, however, for two reasons. First: species that are particularly sensitive to human disturbance and to human-assisted biological invasions are progressively disappearing, leaving the more robust and widespread animals and plants to survive. As we have seen, continental species are surviving relatively well, and most basic types of familiar animals, plants, fungi and microbes are coming through the Anthropocene epoch relatively unscathed. Second: if – and it is a big if – the human population remains relatively stable after the twenty-first century, or declines, and we develop increasingly efficient means of obtaining our food, then the currently increasing human demand for land may go into reverse. 'Peak land use' may be achieved in the next hundred or so years, with pressures progressively reducing thereafter. That should be our goal.

These figures already incorporate the successes of modern conservation efforts, and particularly the protection of national parks and other preserves and actions to save individual endangered species. If these efforts were abandoned, the extinction rate would escalate. A major task of conservation is to keep the losses towards the lower end of the likely range – as well as to encourage biological gains. Although I have been advocating a more flexible approach to the environment, and specifically to conservation, nothing I have said should be used to undermine attempts to save existing species or maintain protected areas.

Yet in this book I have chosen to focus on the gains, which are so often overlooked. Virtually all countries and islands in the world have experienced substantial increases in the numbers of species that can be found in and on them, particularly when it comes to plants. Some of these additional species have colonized human-created habitats under their own volition, and others have been accidentally or deliberately introduced by humans. Most places for which good data are available now have between 20 per cent (for state or country-sized areas within continents) and 100 per cent (especially on islands) more plant species growing in the wild than they did before humans arrived. These percentages may be higher than the global norm, since good data mainly come from countries for which there is a relatively long history of trade, and hence higher rates of importation. However, all parts of the world are now involved in international transport, and these more recently trading countries can be expected to catch up pretty soon. If the recent pattern of more arrivals than disappearances keeps up for another thousand years, most regions will experience at least a doubling of the number of plant species present, and some will see five-fold increases.

Once plant diversity is in place, the diversity of animals that feed on plants, and then of other animals that consume these herbivores, will eventually catch up. Insects are establishing on introduced plants throughout the world, and regular outbreaks of pathogens imply that the same must be happening for fungal and microbial diversity (most such invasions will not be noticed unless they cause a major epidemic). The number of vertebrates in each region is increasing too. Despite the extinction of birds and reptiles on oceanic islands, most of these islands now support more vertebrate species than they used to, if you add up all the introduced mammals, birds, reptiles, freshwater fish and amphibians. The number of species living in virtually every country or island has already increased during the period of human influence, and numbers continue to increase.

These diversity increases go beyond a simple count of the number of species. Each island now supports land mammals as well as birds – the addition of a major new group of animals. Each continental region now contains representatives of plants and animals that evolved for tens of millions of years on separate continents. Adding Australian gum trees to the flora of California increases the variety of major plant groups (*Eucalyptus* is a new genus for California) more than would be achieved by importing an additional pine tree (Monterey pine and other species in the genus *Pinus* already grow there). With this increased diversity at the base of the food chain, there are increased opportunities for insects and fungi and bacteria that did not initially accompany the plants to spread around the world. Almost all countries, states and islands are now more biologically diverse than they used to be.

These ecological changes have added diversity to each region,[4] but they have not added species to the world list – that requires evolution. Evolutionary additions are likely to take place by three main processes. The first is hybridization, as I discussed in Chapter 9. Human-moved species from different parts of the world meet up, reproduce and generate hybrid offspring. In some cases, as with ragworts and monkey-flowers in Britain, the hybrid offspring are not compatible with their parents and new species arise almost as soon as the pollen tube from one species meets the ovule of another. More information is still needed, but the hybrid generation of new plant species in Europe and North America already seems to be as fast as the process of extinction, and possibly even faster. The New Pangea in which distant species are meeting up and generating hybrids is so unusual in the history of life on Earth that the current generation of new plant species, in particular, could be higher than it has been at any time in the 700 million years since there have been land plants. In other cases, the new hybrids (such as between rhododendrons, or between red and sika deer in Scotland) do not form new species immediately, but the new mixed-gene animals or plants will slowly head off in a new evolutionary direction – perhaps generating new species in hundreds of thousands of years, rather than overnight.

The second process involves animals, plants and microbes that take advantage of new ecological situations, as I considered in Chapters 7 and 8. The apple fly and the parasitic insects that attack apple flies became adapted to a new ecological opportunity (introduced apple trees) on a timescale of decades to a century and a half, to the point that they are already on the verge of forming new species. This will be a major process, generating new species not only for decades and centuries but for hundreds of thousands of years to come. A million years hence, we can expect all the plants that establish persistent populations in new parts of the world during the human epoch to have acquired substantial numbers of specialized insects and diseases. The rate of evolution is plenty fast enough to achieve this. Gum tree forests in California are likely to have as many associated insects and diseases in the ten thousandth century as gum trees in Australia have now.

The greatest contribution to new species, however, is likely to be the third process – through geographic separation in the human, and perhaps then post-human, world. This is when two or more populations of the same species, which we have transported around the planet, live in widely separated geographic regions. Eventually, they become distinct species in different places and, just as hybrid speciation is the short-term signature of the Anthropocene, this will be the long-term consequence of the human-dominated Pangean Archipelago. This process of evolutionary divergence in (partial) isolation is how mockingbirds came to be different species on different islands in the Galapagos. It is why, at a larger scale, the South American southern lapwings (the bird that was expressing its vocal irritation at the crowds during the Olympic golf tournaments in Rio de Janeiro) evolved into a species that is different from the related northern lapwing of Europe and Asia, and from other lapwings in Africa, Asia and Australasia. Star-thistles that were introduced from Europe to

California have started off down this road. The Californian plants are already partially infertile when crossed with the pollen of their European ancestors, less than a century after the plants were introduced. In a million years' time, a substantial proportion of all the species that have been transported to new continents and islands will be different enough from their ancestors that they will have become new separate species. By 2 million years AD, almost all of them will be different. Californian and Australian blue gums will no longer be the same species.

To put this in historical context, white-eyes represent a group of tropical birds that spread out across the Old World islands and continents in the last 2 million years.[5] Apart from the mode of transport (flying rather than human-assisted colonization), this is exactly the situation facing every species that has become globally distributed during the Anthropocene. Human-introduced species now live in many different parts of the world, and there are only modest rates of genetic contact today (via continued human movement) between their original populations and the new places where they now live. This is similar to the white-eyes, which needed the capacity to move long distances in the first place but then became adapted to local conditions and became separate species. White-eyes have already diverged into about eighty species.

This is not a one-off. Over 375 species of New World rats and mice evolved after they colonized South America, and rodents have evolved into over 130 species in Australia and New Guinea just a few million years after their first ancestors (presumably) rafted over on floating vegetation from Indonesia. And about eighty species of lupin plants evolved in the Andes in the last one and a half million years, after they invaded from North America. It would take only a modest 5 per cent of the world's species to repeat the white-eye, rodent and lupin's feats (say, generating twenty new species in different geographic locations in a million years) to double the total number of species on our planet. Each of these new species is then likely to acquire unique parasites (including associated insects, in the case of plants) and diseases, further bolstering numbers.

Diversity increases on islands are likely to exceed this. Because introduced mammals can't fly, the populations on each island will become genetically isolated; or at least the rate of continued arrival (gene flow) during the human epoch is unlikely to be sufficient to prevent them from becoming adapted to local conditions. The world's islands will end up with considerably more species in total: tree rats, ground rats, fast rats (where there are predators, they need to escape), slow rats (where there are not), and vegetarian and predatory rats will come into existence in different locations. Give it a few million years after that and many of them will seem truly bizarre. New lineages of reef-diving mammal might be born. Can we put numbers on this? If a thousand of the Pacific Ocean's twenty thousand to thirty thousand islands are large enough to support populations of two kinds of rodent (starting off with one rat and one mouse species) for an extended period of time, we will end up with an enormous number of new species added to the world list. This alone would add, approximately, an extra 40 per cent to the world mammal list. Many of these islands also have cats, dogs, goats, pigs and mongooses, among others, which would also diversify. The authorities in New Zealand have decided that it would be a good idea to wipe out as many introduced mammals as possible but, assuming they fail, there will be dozens of new mammal species that live only there. It seems likely that a third of all mammal diversity might be species that live only on islands, in the year one million AD.

Introduced lizards and birds also abound – there are at least as many introduced bird populations in these scattered locations as there are bird species that have died out. These, too, would turn into separate species, but this time they would be species that are capable of surviving in the presence of mammals. The world's islands will be full of unique vertebrate species again, and probably at least triple the number there were in total before humans arrived (even though there are far fewer unique island species today than there were ten thousand years ago).

Back on the world's continents, we will start to see the largest surviving mammals, including descendants of domestic livestock, reclaiming their place. We can expect the feral horses that already live in North America, South America, Europe and Australia to become separate species in a million years or so; and there might be more than one species on each continent. Descendants of tapirs may be on their way to becoming the next multi-tonne beasts. Giant jaguars will have returned to prey on them. However, these new biological communities will probably be dominated by even-toed ungulates. Relatives of cattle, goats and fleet-of-foot deer and antelope have survived the human epoch pretty well. These are the mammals that will be available to start evolving into the next generation of megabeasts. However, there may be unexpected additions. The fastest kangaroos, if they escape from captivity during the human epoch, could by then have become major herbivores in all the world's grasslands and savannas.

Having turned the whole world into a global archipelago, we have set in train processes that will increase rather than decrease the long-term diversity of the Earth. If I were to hazard a wild guess, and humans were to disappear today, I imagine that there will be approximately double (between one and a half and five times) the number of species alive on the Earth in a million years. If the current rates at which species are accumulating in new parts of the world were to continue for another millennium, and then humans were to disappear quietly (without a major calamity), then the multiplication of diversity over the course of the next million years could be far higher than this. Five million years later, once these new species have evolved into increasingly distinct forms, the advent of humans could be seen to have substantially increased the biological diversity of the Earth. A sixth genesis.

Acknowledgements

The bulk of the credit for everything in this book goes to those scientists, conservationists and many others I have met throughout my life, and who have influenced my thoughts. I would also like to thank dozens of others who accompanied me, hosted me, argued with me and made various trips around the world the enjoyable times they were. I am not going to try to thank everyone individually – but thank you. I would like to thank all the authors of the scientific articles and books I have read on the influences of humans on the biological world, and those who have studied how species and biological communities have responded to climatic and other environmental changes in the geological past – this historical context is essential if we are to understand and then respond to present-day changes. I cite some of these works directly, especially where I dwell on particular examples. However, thousands of articles have been consulted (and tens of thousands over the years), hence it is impossible to list them all without making the Notes longer than the whole of the rest of the book. My particular thanks to all those whose work I should have cited but have not.

I have done my best to represent all factual information accurately, even when my own interpretation of the evidence is not always the same as that of the original authors. All the opinions represented in this book remain my own, although I have of course borrowed liberally from others. I am as grateful to those who have argued against my ideas as to those who have agreed – honest discussion is how science advances. I apologize to those who feel offended by my conclusions. I hope that we can shake hands and productively discuss the best way forward.

In a couple of places, I have included a new analysis in this book (such as the effects of sparrows on bluebirds). In these circumstances, I have provided a very brief summary in the text or in endnotes to the relevant chapters.

All the events described in this book are accurate to the best of my recollection and knowledge; for the purposes of the narrative, some things that happened or were observed up to two years apart are presented as though they took place as a single event. I have personally visited most of the locations described in this book, but on a few occasions I have used accounts by others (including information in the methods sections of scientific articles) and online photographs in an attempt to bring places and species 'to life'. Where I have done so, I have taken in good faith the information that is publicly available. The Leverhulme Trust and the Georgetown Environment Initiative supported the Chernobyl trip; thanks to Victoria Beale and David Moon for organizing it.

A number of people have kindly read draft chapters and sections, or otherwise provided information, in an attempt to spot any factual errors or misinterpretations I may have inadvertently made. They are: Richard Bailey, Colin Beale, Jon Bridle, Roger Butlin, Matthew Collins, Erle Ellis, Jane Hill, Daniel Montesinos, Dov Sax and Stuart Weiss. Very many thanks. Any errors remain my own.

I thank my agent, Peter Tallack, for persuading me to write the book in the first place, and Peter and Tisse Takagi for their wise advice throughout the process. Many thanks to my book editors Laura Stickney (Penguin Random House) and Ben Adams (Public Affairs Books), and Sarah Day, for their suggestions and valiant attempts to convert my prose into a readable volume. I hope it was not too painful. I also thank Shoaib Rokadiya, Pen Vogler (publicity) and Richard Duguid (editorial) at Penguin, and Melissa Raymond (production), Jaime Leifer (publicity) and Lindsay Fradkoff (marketing) at Public Affairs Books, and their colleagues.

Thanks to my wife, Helen, daughters Rose, Alice and Lucy, and son, Jack, and for everyone at work for tolerating my lack of attention to my 'real job' while I was writing this book. My sisters, Phil, Julie and Anthy, and brother, Jeremy, helped recall anecdotes from my youth and found old family photos; they and my brothers-in-law,

ACKNOWLEDGEMENTS

Steve and Erik, and my sister-in-law, Sarah, provided information and photos relating to one or more of the trips described in this book.

Then to the disappointments. I really did not need to be asked quite so often: 'How is your book going?' You know who you are. Now you can judge for yourselves.

Picture Acknowledgements

All photographs by the author, unless specified – many thanks to the other photographers. Copyright of photographic material remains with the original photographer.

CHAPTER 1

Photo: House sparrow copyright © Kevin Phillips.

CHAPTER 2

Maps of megafauna diversity: Thanks to Søren Faurby and Jens-Christian Svenning for providing the data for present and potential diversity, and to Phil Platts, who kindly produced the maps.
Photo: Lucy (child) and Rex (wolfhound), copyright © Aldina Franco.
Photo: Family wedding in 1951, probably by Ambler Thomas.

CHAPTER 4

Photo: Giant mole-rat copyright © Stuart Orford.
Comma butterfly map: thanks to Georgina Palmer.

CHAPTER 6

Photo: Pukeko copyright © Tony Wills.
Photo: Takahe copyright © Ashleigh Thompson.

CHAPTER 7

Photos: Edith's checkerspot butterfly number 6 copyright © Michael C. Singer; remaining checkerspot adults, eggs and larvae copyright © Paul Stevens.

Photo: Chris Thomas in a snowy campsite copyright © Michael C. Singer and Camille Parmesan.

Bulldog images:

1. *Bulldog* (1790) by Philip Reinagle.
2. *Handsome Dan* (around 1890), photograph by Pach Brothers (courtesy of Yale University Manuscripts & Archives Digital Images Database, public domain).
3. Standing bulldog (2010) copyright © Wikipedia Creative Commons.

CHAPTER 8

Photos: *Diamante* boat/seascape, and mockingbird portrait, copyright © Stephen Dempsey; adult and young mockingbird copyright © Erik Vikander.

Photo: Californian yellow star-thistle copyright © Daniel Montesinos.

Photos: Hawthorn fly, apple fly and parasitoid copyright © Hannes Schuler.

CHAPTER 9

Photo: Kakapo copyright © Andrew Digby (Department of Conservation).

Photo: Jacqueline Beggs; courtesy of Jacqueline Beggs.

CHAPTER 11

Photos: Monarch butterfly and *Eucalyptus* copyright © Stuart Weiss.

Photos: Monterey pine photographs copyright © Chris Earle.

Notes

PROLOGUE

1. I am defining *ecological success* as survival in human-altered locations, and *ecological gain* to be an increase in the numbers of individuals, range of habitats or geographic distribution of a particular species, irrespective of whether any evolutionary change has taken place. I also include increases in diversity in a given location within my definition of ecological gain. I consider *evolutionary success* and *gain* to be instances where survival (when considering success) and increased abundance, use of new habitats and an enlarged distribution (when considering gains) are underpinned by evolutionary change or by the previously evolved characteristics of different types of species. I include evolved increases in diversity (such as one species evolving into two) within my definition of evolutionary gain.
2. The IUCN Red List of Threatened Species summary statistics, http://www.iucnredlist.org/about/summary-statistics#How_many_threatened, accessed 1 January 2017.
3. I leave others to discuss how many of my conclusions hold for the marine realm.

CHAPTER 1: BIOGENESIS

1. Sætre, G. P. et al. (2012), 'Single origin of human commensalism in the house sparrow', *Journal of Evolutionary Biology*, 25, 788–96.
2. Lever, C. (2005), *Naturalised Birds of the World*, London: Poyser.
3. Robinson, R. A. et al. (2015), *BirdTrends 2015: Trends in numbers, breeding success and survival for UK breeding birds*. Research Report 678, Thetford: BTO; Rich, T. D. et al. (2004), *Partners in Flight: North American landbird conservation plan*, Ithaca, NY: Cornell Laboratory

of Ornithology; BirdLife International (2016) Species factsheet: *Passer domesticus*, downloaded from http://www.birdlife.org, 12 January 2016.

4. No criticism of the research itself. I am merely questioning why the decline in the house-sparrow population is regarded as an ecological or conservation crisis. Research on this topic includes: Hole, D. G. et al. (2002), 'Agriculture: Widespread local house-sparrow extinctions', *Nature*, 418, 931–2; Peach, W. J., Sheehan, D. K. & Kirby, W. B. (2014), 'Supplementary feeding of mealworms enhances reproductive success in garden nesting House Sparrows *Passer domesticus*', *Bird Study*, 61, 378–85.

5. Dawson, W. L. (1903), *Birds of Ohio*, Columbus, Ohio: Wheaton Publishing Company.

6. http://michiganbluebirds.org, accessed 29 July 2015.

7. Analysis of Audubon Christmas Bird Count data 1951–2014. Sparrow numbers have declined by over 60 per cent (an average of 12.16 house sparrows per party hour were counted in 1951–60, compared to 4.51 per party hour in 2005–14), while Eastern bluebirds have increased by about 50 per cent (an average of 0.42 bluebirds per party hour in 1951–60, compared to 0.62 per party hour in 2005–14). Eastern bluebirds do not decline (change in counts from one year to the next) following years when the actual numbers of sparrows is high (R^2 = 0.0014, n = 63 years, p = 0.77), or when numbers of sparrows increase (R^2 = 0.0031, n = 62 years, p = 0.67). Equally, increases and decreases in sparrow numbers are not related to the numbers of bluebirds or to changes in the numbers of bluebirds. The North American Breeding Bird Survey (BBS), which measures abundances during the breeding season, also finds a statistically significant increase in the overall abundance of Eastern bluebirds and a decline in the numbers of house sparrows (over the period 1966–2013), including in the state of Michigan.

8. Hall, K. D. (2007), 'Guidelines for successful monitoring of Eastern Bluebird nest boxes', *Passenger Pigeon*, 69, issue 2.

9. Peterson, R. T., Mountfort, G. R. & Hollum, P. A. D. (1966), *A Field Guide to the Birds of Britain and Europe*, London: Collins (revised edition).

10. Thanks to Fabrice Eroukhmanoff and Anna Runemark, as well as to Glenn-Peter Sætre and Richard Bailey. Research students in the project include Jo Hermansen, Tore Elgvin and Cassandra Trier.

11. Hermansen, J. S. et al. (2011), 'Hybrid speciation in sparrows I: Phenotypic intermediacy, genetic admixture and barriers to gene flow', *Molecular Ecology*, 20, 3812–22; Elgvin, T. O. et al. (2011), 'Hybrid speciation in

sparrows II: A role for sex chromosomes?', *Molecular Ecology*, 20, 3823–37; Trier, C. N. et al. (2014), 'Evidence for mito-nuclear and sex-linked reproductive barriers between the hybrid Italian sparrow and its parent species', *PLOS Genetics*, 10, e1004075.

12. Zeder, M. A. (2008), 'Domestication and early agriculture in the Mediterranean Basin: Origins, diffusion, and impact', *Proceedings of the National Academy of Sciences USA*, 105, 11597–604.

13. Based on a twenty-minute internet search of current news; stories dated 3–17 December 2016.

14. CNN quote; http://edition.cnn.com/2016/12/12/world/sutter-vanishing-help/.

15. Henderson, I. S. (2010), 'North American Ruddy Ducks *Oxyura jamaicensis* in the United Kingdom – population development and control', *BOU Proceedings – The Impacts of Non-native Species*.

16. BirdLife International (2015), *Himantopus novaezelandiae*, IUCN Red List of Threatened Species, 2015.

17. For readers unfamiliar with rugby, the New Zealand rugby team is their greatest source of sporting pride, and the team is known as the 'All Blacks'.

CHAPTER 2: FALL AND RISE

1. The 30 December 2016 Chinese Government announcement of a cessation of commercial ivory sales and processing by the end of 2017 will, hopefully, reduce this problem; http://www.gov.cn/zhengce/content/2016-12/30/content_5155017.htm

2. Wittemyer, G. et al. (2014), 'Illegal killing for ivory drives global decline in African elephants', *Proceedings of the National Academy of Sciences USA*, 111, 13117–21.

3. Less so in West Africa, where cattle show some degree of resistance.

4. Some suggest that African elephants are three, rather than two, species, in which case four survive in total.

5. Climatic changes also contributed, but only in the sense that humans were more likely to exterminate a species at times when their numbers were reduced by the climate; Cooper, A. et al. (2015), 'Abrupt warming events drove Late Pleistocene Holarctic megafaunal turnover', *Science*, 349, 602–6.

6. Fariña, R. A, Vizcaíno, S. F. & De Iuliis, G. (2013), *Megafauna: Giant Beasts of Pleistocene South America*, Bloomington, Indiana: Indiana University Press.

7. Based on average weights of males and females. A few bull eland and dromedaries weigh over a tonne, but their average is lower. The giraffe has recently been divided into four separate species; Fennessy, J. et al. (2016), 'Multi-locus analyses reveal four giraffe species instead of one', *Current Biology*, 26, 2543–9.

8. Redrawn using data from Faurby, S. & Svenning, J.-C. (2015), 'Historic and prehistoric human-driven extinctions have reshaped global mammal diversity patterns', *Diversity and Distributions* 21, 1155–66.

9. Duncan, R. P., Boyer, A. G. & Blackburn, T. M. (2013), 'Magnitude and variation of prehistoric bird extinctions in the Pacific', *Proceedings of the National Academy of Sciences USA*, 110, 6436–41.

10. These estimates include species listed by IUCN as (a) extinct in the wild (for which live specimens still survive in zoos, botanic gardens or seed banks), (b) completely extinct, and (c) possibly extinct and possibly extinct in the wild. This is the best estimate currently available.

11. I am referring to the very large four-legged (and two-legged, two-handed) dinosaurs; birds are flying dinosaurs.

12. The genetic (DNA) sequences of species diverge with increasing time, and genetic differences can be calibrated using fossils of known age. Authors are not in full agreement over the timing of events, but much bird diversity did exist before the terrestrial dinosaurs disappeared. Jetz, W. et al. (2012), 'The global diversity of birds in space and time', *Nature*, 491, 444–8; Claramunt, S. & Cracraft, J. (2015), 'A new time tree reveals Earth history's imprint on the evolution of modern birds,' *Science Advances*, 1, e1501005.

13. Moyle R. G. et al. (2016), 'Tectonic collision and uplift of Wallacea triggered the global songbird radiation', *Nature Communications* 7, 12709.

14. Bininda-Emonds, O. R. P. et al. (2007), 'The delayed rise of present-day mammals', *Nature*, 446, 507–12; Meredith, R. W. et al. (2011), 'Impacts of the Cretaceous Terrestrial Revolution and KPg extinction on mammal diversification', *Science*, 334, 521–4; O'Leary, M. A. et al. (2013), 'The placental mammal ancestor and the post-'K-Pg radiation of placentals', *Science*, 339, 662–7; Puttick, M. N., Thomas, G. H. & Benton, M. J. (2016), 'Dating placentalia: Morphological clocks fail to close the molecular fossil gap', *Evolution*, 70, 873–6.

15. Rainford, J. L. et al. (2014), 'Phylogenetic distribution of extant richness suggests metamorphosis is a key innovation driving diversification in insects', *PLOS ONE*, 9, e109085.

16. Nichols, D. J. & Johnson, K. R. (2008), *Plants and the KT Boundary*. Cambridge: Cambridge University Press.

17. Evans, A. R., et al. (2012), 'The maximum rate of mammal evolution', *Proceedings of the National Academy of Sciences USA*, 109, 4187–90.

18. McGovern, P. E. (2003), *Ancient Wine: The Search for the Origins of Viniculture*, Princeton, New Jersey: Princeton University Press.

19. UN FAO figures for 2013.

20. Smil, V. (2011), 'Harvesting the biosphere: The human impact', *Population and Development Review*, 613–36; Smith, F. A. et al. (2016), 'Megafauna in the Earth system', *Ecography*, 39, 99–108.

21. Barnosky, A. D. (2008), 'Megafauna biomass tradeoff as a driver of Quaternary and future extinctions', *Proceedings of the National Academy of Sciences USA*, 105 (Supplement 1), 11543–8; this statement does not apply to marine mammals, where numbers of the heaviest species collapsed far more recently and are only now recovering.

22. Herrero, M. et al. (2013), 'Biomass use, production, feed efficiencies, and greenhouse gas emissions from global livestock systems', *Proceedings of the National Academy of Sciences USA*, 110, 20888–93.

23. On land only. Hunting (with sonars and nets) is still the norm in the oceans.

24. Wearing fur remains sensible when it is a by-product of farming animals for their meat (it avoids waste). However, the modern stance against fur coats extends to social condemnation of wearing fur from our livestock. This condemnation does not generally extend to their skin (leather for shoes, etc.).

25. For example, fur farming and animal coats remain popular in parts of Asia.

26. Deinet, S. et al. (2013), *Wildlife Comeback in Europe: The Recovery of Selected Mammal and Bird Species*, London: Zoological Society of London.

27. The present-day elephants may be derived from more than one introduction to Borneo because elephants were common gifts between rulers and have been transported by sea in the region for over six hundred years; Cranbrook, E., Payne, J. & Leh, C. M. U. (2008), 'Origin of the elephants *Elephas maximus* of Borneo', *Sarawak Museum Journal*, 63, 1–25.

28. Shim, P. S. (2003), 'Another look at the Borneo elephant', *Sabah Society Journal*, 20, 7–14.

29. One of the 'Aichi targets' for 2020 (Convention on Biological Diversity Strategic Plan for Biodiversity) relates to invasive alien species. It includes the control or eradication of priority species.

30. Fernando, P. et al. (2003), 'DNA analysis indicates that Asian elephants are native to Borneo and are therefore a high priority for conservation,' *PLOS Biology*, 1, 110–15. These authors did not consider the alternative and more likely scenario that they were introduced from a genetically unique population elsewhere (probably Java). Apparent genetic uniqueness could also arise if the existing population is the product of multiple introductions from different locations.

CHAPTER 3: NEVER HAD IT SO GOOD

1. Corrales, F. & Badilla, A. (2005), *Investigaciones Arqueologicas en Sitios con Esferas de Piedra*, San José, Costa Rica: Delta del Diquís.
2. Roberts, A. (2011), *Evolution: The human story*, London: Dorling Kindersley; Ellis, E. C. et al. (2013), 'Used planet: A global history', *Proceedings of the National Academy of Sciences*, 110, 7978–85.
3. Galetti, M. et al. (2009), 'Hyper abundant mesopredators and bird extinction in an Atlantic forest island', *Zoologia*, 26, 288–98.
4. Based on the species–area relationship, which commonly takes the form $\log S = \log C + Z^*\log A$, where the number of species is represented by S and the area of an island or habitat remnant is denoted A, and C and Z are constants. The two estimates of survival given are based on $Z = 0.15$ (which is typical of samples from different areas on land) and $Z = 0.25$ (which is commonly the case when the samples are from water-surrounded islands – and would be appropriate for forest specialists that were isolated from other forests).
5. Brooks, T. & Balmford, A. (1996), 'Atlantic forest extinctions', *Nature*, 380, 115. The authors split the analysis into four sub-regions, and then compiled the estimates of extinction for each region into one overall value.
6. Brooks, T., Tobias, J. & Balmford, A. (1999), 'Deforestation and bird extinctions in the Atlantic forest', *Animal Conservation*, 2, 211–22.
7. The situation is potentially even worse because the remaining forest is subdivided into small fragments; Schnell, J. K. et al. (2013), 'Quantitative analysis of forest fragmentation in the Atlantic Forest reveals more threatened bird species than the current Red List', *PLOS ONE*, 8, e65357.
8. Canale, G. R. et al. (2012), 'Pervasive defaunation of forest remnants in a tropical biodiversity hotspot', *PLOS ONE*, 7, e41671.
9. Gonçalves da Cruz, C. A. & Pimenta, B. (2004), *Phrynomedusa fimbriata*, IUCN Red List of Threatened Species, version 2014.2: www.iucnredlist.org.

10. Vellend, M. et al. (2013), 'Global meta-analysis reveals no net change in local-scale plant biodiversity over time', *Proceedings of the National Academy of Sciences USA*, 110, 19456–9; Dornelas, M. et al. (2014), 'Assemblage time series reveal biodiversity change but not systematic loss', *Science*, 344, 296–9.

11. Newbold, T. et al. (2015), 'Global effects of land use on local terrestrial biodiversity', *Nature*, 520, 45–50. Estimated losses in this article relate to the number of species per unit area (e.g. per square metre or per hectare) of a single habitat, and do not refer to the number of species in an entire landscape.

12. Matthias, W. et al. (2005), 'From forest to farmland: Habitat effects on Afrotropical forest bird diversity', *Ecological Applications*, 15, 1351–66.

13. Bobo, K. S. et al. (2006), 'From forest to farmland: Butterfly diversity and habitat associations along a gradient of forest conversion in south-western Cameroon', *Journal of Insect Conservation*, 10, 29–42.

14. Waltert, M. et al. (2011), 'Assessing conservation values: Biodiversity and endemicity in tropical land use systems,' *PLOS ONE*, 6, e16238.

15. Maes, D. & Van Dyck, H. (2001), 'Butterfly diversity loss in Flanders (north Belgium): Europe's worst case scenario?', *Biological Conservation*, 99, 263–76.

16. Many recent losses were associated with historical habitats that were created by humans in the first place.

17. Hanski, I. (2016), *Messages from Islands: A Global Biodiversity Tour.* Chicago: University of Chicago Press.

18. Rosenzweig, M. L. (1995), *Species Diversity in Space and Time,* Cambridge: Cambridge University Press.

19. Hiley, J. R., Bradbury, R. B. and Thomas, C. D. (2016), 'Impacts of habitat change and protected areas on alpha and beta diversity of Mexican birds', *Diversity and Distributions*, 22, 1245–54.

20. Stein, A., Gerstner, K. & Kreft, H. (2014), 'Environmental heterogeneity as a universal driver of species richness across taxa, biomes and spatial scales', *Ecology Letters*, 17, 866–80.

CHAPTER 4: STEAMING AHEAD

1. Gippoliti, S. & Hunter, C. (2008), *Theropithecus gelada*, IUCN Red List of Threatened Species 2008, downloaded 27 March 2016.

2. The air cools at approximately 0.5°C per 100m of increased elevation when the air is saturated with water, but nearer 1°C per 100m in dry air.

3. This is for Scenario RCP 8.5 (warming may not be this great if greenhouse gas controls are more effective in future). Barros, V. R. et al. (2015), *Climate Change 2014: Impacts, Adaptation, and Vulnerability. Part B: Regional Aspects,* Contribution of Working Group II to the Fifth Assessment Report of the Intergovernmental Panel on Climate Change.

4. Marino, J. & Sillero-Zubiri, C. (2013), *Canis simensis,* IUCN Red List of Threatened Species 2013, downloaded 27 March 2016.

5. Sillero-Zubiri, C., Tattersall, F. H. & Macdonald, D. W. (1995), 'Habitat selection and daily activity of giant mole-rats (*Tachyoryctes macrocephalus*): Significance to the Ethiopian wolf (*Canis simensis*) in the Afroalpine ecosystem', *Biological Conservation* 72, 77–84.

6. The Bale Mountains National Park may be able to prevent this kind of encroachment.

7. The land that is any higher is too steep and rocky to support large numbers of burrowing giant mole-rats; and the area is much reduced.

8. Chen, I-C. et al. (2009), 'Elevation increases in moth assemblages over 42 years on a tropical mountain', *Proceedings of the National Academy of Sciences USA*, 106, 1479–83; Chen, I-C. et al. (2011), 'Asymmetric boundary shifts of tropical montane Lepidoptera over four decades of climate warming', *Global Ecology and Biogeography*, 20, 34–45.

9. Pounds, J. A. et al. (2006), 'Widespread amphibian extinctions from epidemic disease driven by global warming', *Nature*, 439, 161–7. Most of the extinctions coincide with El Niño currents in the Pacific Ocean, altering temperatures, cloud cover and rainfall in the New World tropics. These hot years are superimposed on the background warming generated by human-caused climate change, so each new 'peak' is higher than the last (e.g. record-breaking 1997–8 El Niño peak temperatures, when many harlequin frogs disappeared, have now been exceeded by 2015–16 El Niño temperatures).

10. Thomas, C. D. et al. (2004), 'Extinction risk from climate change', *Nature*, 427, 145–8.

11. Stopping climate change is the top priority. However, so much climate change has already taken place and is certain to occur in future (e.g. because of existing power stations) that implementing measures to move populations is already the only realistic option for many species.

12. Coope, G. R. (1979), 'Late Cenozoic fossil Coleoptera: Evolution, biogeography, and ecology', *Annual Review of Ecology and Systematics*, 10, 247–67.

13. Divers are called 'loons' in North America.

14. 'Alluvial Archaeology in the Vale of York. The Geology of the Vale of York', http://www.yorkarchaeology.co.uk/valeofyork/geology.htm; and a presentation by Paul Buckland, http://www.thmcf.org/downloads/ Humberhead%20Levels%20Rise%20and%20Fall.pdf.

15. Muscheler, R. et al. (2008), 'Tree rings and ice cores reveal 14C calibration uncertainties during the Younger Dryas', *Nature Geoscience*, 1, 263–7.

16. Human artefacts are known from Creswell Crags in the adjacent county of Derbyshire at this time.

17. Clark, P. U. et al. (2009), 'The Last Glacial Maximum', *Science*, 325, 710–14.

18. Mayle, F. E. et al. (2004), 'Responses of Amazonian ecosystems to climatic and atmospheric carbon dioxide changes since the last glacial maximum', *Philosophical Transactions of the Royal Society, B.*, 359, 499–514; Morley, R. J. (2000), *Origin and Evolution of Tropical Rain Forests*, New York: John Wiley & Sons.

19. Kuper, R. & Kröpelin, S. (2006), 'Climate-controlled Holocene occupation in the Sahara: Motor of Africa's evolution', *Science*, 313, 803–7.

20. Smith, S. E. et al. (2013), 'The past, present and potential future distributions of cold-adapted bird species', *Diversity and Distributions*, 19, 352–62.

21. United Kingdom's National Biodiversity Gateway: https://data.nbn. org.uk/.

22. Heath, J., Pollard, E. & Thomas, J. A. (1984), *Atlas of Butterflies in Britain and Ireland*, Harmondsworth: Viking. The one 1970–82 record from Scotland is thought to be a stray individual that had not established a breeding population.

23. Braschler, B. & Hill, J. K. (2007), 'Role of larval host plants in the climate-driven range expansion of the butterfly *Polygonia c-album*', *Journal of Animal Ecology*, 76, 415–23. Comma caterpillars also eat wych elm, but these trees are rarer than nettles and have declined, whereas nettles have increased.

24. Thomas, C. D. (2010), 'Climate, climate change and range boundaries', *Diversity and Distributions*, 16, 488–95; Illán, J. G. et al. (2014), 'Precipitation and winter temperature predict long-term range-scale abundance changes in Western North American birds', *Global Change Biology*, 20, 3351–64.

25. Chen, I-C. et al. (2011), 'Rapid range shifts of species associated with high levels of climate warming', *Science*, 333, 1024–6. This rate of

movement is the median; some are moving much faster and others are barely moving.

26. Boyd, P. W. et al. (2013), 'Marine phytoplankton temperature versus growth responses from polar to tropical waters – outcome of a scientific community-wide study', *PLOS ONE*, 8, p.e63091.

27. Average annual temperature in Britain varies between 8.5°C (in the north) and 11°C (in the south).

28. Hawkins, B. A. et al. (2003), 'Energy, water, and broad-scale geographic patterns of species richness', *Ecology*, 84, 3105–17.

29. Menéndez, R. et al. (2006), 'Species richness changes lag behind climate change', *Proceedings of the Royal Society of London B*, 273, 1465–70. These climate-related diversity increases have taken place despite declines of many individual species, associated with changes to farming and forestry practices.

30. Pounds, J. A., Fogden, M. P. & Campbell, J. H. (1999), 'Biological response to climate change on a tropical mountain,' *Nature*, 398, 611–15.

31. Sommer, J. H. et al. (2010), 'Projected impacts of climate change on regional capacities for global plant species richness', *Proceedings of the Royal Society of London B*, 277, 2271–80; Reu, B. et al. (2011), 'The role of plant functional trade-offs for biodiversity changes and biome shifts under scenarios of global climatic change', *Biogeosciences*, 8, 1255–66; Venevskaia, I., Venevsky, S. & Thomas, C. D. (2013), 'Projected latitudinal and regional changes in vascular plant diversity through climate change: Short-term gains and longer-term losses', *Biodiversity and Conservation*, 22, 1467–83.

32. Mayhew, P. J. et al. (2012), 'Biodiversity tracks temperature over time', *Proceedings of the National Academy of Sciences USA*, 109, 15141–5.

33. Wilson, R. J. et al. (2007), 'An elevational shift in butterfly species richness and composition accompanying recent climate change', *Global Change Biology*, 13, 1873–7.

34. Predictions for the world's drylands are complicated because: there is uncertainty about local and regional patterns of future rainfall; increased evaporation at higher temperatures results in some drying, even if rainfall increases slightly; the soil affects water retention; and the increased concentration of carbon dioxide in the atmosphere allows plants to keep the stomatal pores in their leaves closed for more of the time (reducing moisture loss). There is also uncertainty about what will happen to diversity in the world's hottest rainforests (because the future climate in these locations will be unlike any that currently exist).

CHAPTER 5: PANGEA REUNITED

1. Hofstetter, S. et al. (2006), 'Late glacial and Holocene vegetation history in the Insubrian Southern Alps – new indications from a small-scale site', *Vegetation History and Archaeobotany*, 15, 87–98.

2. Sandom, C. J. et al. (2014), 'High herbivore density associated with vegetation diversity in interglacial ecosystems', *Proceedings of the National Academy of Sciences USA*, 111, 4162–7.

3. Evans, K. L. et al. (2010), 'A conceptual framework for the colonisation of urban areas: The blackbird *Turdus merula* as a case study', *Biological Reviews*, 85, 643–67.

4. Walther, G.-R. et al. (2002), 'Ecological responses to recent climate change', *Nature*, 416, 389–95.

5. Volta, P. & Jepsen, N. (2008), 'The recent invasion of *Rutilus rutilus* (L.) (Pisces: Cyprinidae) in a large South-Alpine lake: Lago Maggiore', *Journal of Limnology*, 67, 163–70.

6. Hobbs, R. J. et al. (2006), 'Novel ecosystems: Theoretical and management aspects of the new ecological world order', *Global Ecology & Biogeography* 15, 1–7; Ellis, E. C. & Ramankutty, N. (2008), 'Putting people in the map: Anthropogenic biomes of the world', *Frontiers in Ecology and the Environment*, 6, 439–47.

7. It is not always possible to distinguish between closely related species on the basis of fossil remains, so it is more reliable to count numbers of plant genera (plural of 'genus') than numbers of separate species. For example, different species of oak tree in the genus *Quercus* can have pollen grains that are too alike to distinguish easily.

8. Svenning, J. C. (2003), 'Deterministic Plio-Pleistocene extinctions in the European cool-temperate tree flora', *Ecology Letters*, 6, 646–53.

9. Russell, J. C. & Blackburn, T. M. (2016), 'The rise of invasive species denialism', *Trends in Ecology & Evolution* DOI 10.1016/j.tree.2016.10.012. The debate arises because defining a species as having 'negative impact' is a human construct. Scientists can agree on the facts but still differ in their interpretation: for instance, the arrival of evergreens around Maggiore increases the total diversity of the region (positive impact) but decreases the abundance of some of the deciduous trees that were there previously (negative impact).

10. Nielsen, S. V. et al. (2011), 'New Zealand geckos (Diplodactylidae): Cryptic diversity in a post-Gondwanan lineage with trans-Tasman affinities', *Molecular Phylogenetics & Evolution*, 59, 1–22; England, R. (2013), *New Zealand passerines: A contribution to passerine phylogeny'*, M.Sc.

thesis, Palmerston North, New Zealand: Massey University; Mitchell, K. J. et al. (2014), 'Ancient DNA reveals elephant birds and kiwi are sister taxa and clarifies ratite bird evolution', *Science*, 344, 898–900; Carr, L. M. et al. (2015), 'Analyses of the mitochondrial genome of *Leiopelma hochstetteri* argues against the full drowning of New Zealand', *Journal of Biogeography*, 42, 1066–76.

11. Dyer, E. E. et al. (2017), 'The global distribution and drivers of alien bird species richness', *PLOS Biology*, 15, p.e2000942.

12. http://www.europe-aliens.org/default.do; trade is increasingly global, so Europe is probably no more than a century ahead of other regions.

13. Addison, D. J. & Matisoo-Smith, E. (2010), 'Rethinking Polynesians' origins: A West-Polynesia triple-I model', *Archaeology in Oceania*, 45, 1–12.

14. Duncan, R. P., Boyer, A. G. & Blackburn, T. M. (2013), 'Magnitude and variation of prehistoric bird extinctions in the Pacific', *Proceedings of the National Academy of Sciences USA*, 110, 6436–41.

15. Nuwer, R. (2013), 'Doctors used to use live African frogs as pregnancy tests', smithsonian.com.

16. Epidemics take place in hot years, as described in Chapter 4.

17. Young, H. S. et al. (2016), 'Patterns, causes, and consequences of Anthropocene defaunation', *Annual Review of Ecology, Evolution, and Systematics*, 47, 333–58.

18. Most of the 'damage' attributed to invasive species on continents involves changes in the abundances, habitats and distributions of 'native' species, rather than the extinction of species. My emphasis is on whether entire species become extinct because there is no prospect that we can (or should aim to) maintain the exact status quo of each species in the face of climate change, habitat change, nitrogen deposition and invasions.

19. Chestnut blight is a disease that did not (quite) extinguish the American chestnut, but several species of plant-feeding insect that used to rely on this tree are thought to have become extinct.

20. Paine, R. T. (1969), 'A note on trophic complexity and community stability', *American Naturalist*, 103, 91–3.

21. Williamson, M. (1996), *Biological Invasions*, London: Chapman & Hall. Williamson recognizes that the percentage of species establishing at each stage varies, for example differing between animals and plants, and between continents and islands. It can be rather higher on isolated islands.

22. Thomas, C. D. & Palmer, G. (2015), 'Non-native plants add to the British flora without negative consequences for native diversity', *Proceedings of the National Academy of Sciences USA*, 112, 4387–92.

23. Kauri dieback disease (an oomycete mould) threatens the survival of the most statuesque of the forest trees. This is a genuine concern that demands urgent action.

24. Vellend, M. et al. (2016), 'Plant biodiversity change across scales during the Anthropocene', *Annual Review of Plant Biology*, 68. DOI 10.1146 /annurev-arplant-042916-040949.

25. Clayson, J. et al. (2006), *New Zealand Naturalised Vascular Plant Checklist*, Wellington: New Zealand Plant Conservation Network.

26. Sax, D. F. and Gaines, S. D. (2008), 'Species invasions and extinction: The future of native biodiversity on islands', *Proceedings of the National Academy of Sciences USA*, 105 (Supplement 1), 11490–97.

27. Sax, D. F. & Gaines, S. D. (2003), 'Species diversity: From global decreases to local increases', *Trends in Ecology & Evolution*, 18, 561–6.

28. Thomas, C. D. (2013), 'The Anthropocene could raise biological diversity', *Nature*, 502, 7; Roy, H. E. et al. (2012), *Non-Native Species in Great Britain: Establishment, Detection and Reporting to Inform Effective Decision Making*, London: Department for Environment, Food and Rural Affairs.

29. Ellis, E. C., Antill, E. C. & Kreft, H. (2012), 'All is not loss: Plant biodiversity in the Anthropocene', *PlOS ONE*, 7, p.e30535.

30. Some British species have declined as a result of the arrival of foreign species, and a few (such as the red squirrel) have died out from parts of their British ranges; but they have not disappeared completely. All the invasion-threatened species in Britain still occur elsewhere in continental Europe.

CHAPTER 6: HEIRS TO THE WORLD

1. Durrell, G. (1966), *Two in the Bush,* London: Collins.

2. At the time of writing in 2016, the New Zealand government has set aside NZ$28m, which is less than a third of 1 per cent of the estimated cost. The project is not necessarily impossible, if new technologies can be developed. However, reinvasion will be a continuous threat for as long as transport continues.

3. http://www.rationaloptimist.com/blog/eradicating-rats-from-oceanic-islands/; http://www.bbc.co.uk/news/uk-scotland-tayside-central-3327 6540.

4. For example as an experiment on predator-inhabited islands and in large enclosures on the mainland to find out how New Zealand's surviving animals and plants respond to pedestrian birds.

5. Pearce, F. (2015), *The New Wild: Why Invasive Species Will be Nature's Salvation*, Boston, MA: Beacon Press.

6. Carleton, M. D. & Musser, G. G. (2005), 'Order Rodentia', in Wilson, D. E. & Reeder, D. M. (eds.), *Mammal Species of the World: A Taxonomic and Geographic Reference*, Volume 12, pp. 745–52, Baltimore: Johns Hopkins University Press

7. Hand, S. J. et al. (2009), 'Bats that walk: A new evolutionary hypothesis for the terrestrial behaviour of New Zealand's endemic mystacinids', *BMC Evolutionary Biology*, 9, 169.

8. Simmons, N. B. et al. (2008), 'Primitive Early Eocene bat from Wyoming and the evolution of flight and echolocation', *Nature*, 451, 818–21; Bininda-Emonds, O. R. P. et al. (2007), 'The delayed rise of present-day mammals', *Nature*, 446, 507–12; Meredith, R. W. et al. (2011), 'Impacts of the Cretaceous Terrestrial Revolution and KPg extinction on mammal diversification', *Science*, 334, 521–4.

9. Jetz, W. et al. (2012), 'The global diversity of birds in space and time', *Nature*, 491, 444–8; Claramunt, S. & Cracraft, J. (2015), 'A new time tree reveals Earth history's imprint on the evolution of modern birds', *Science Advances*, 1, e1501005.

10. Wright, N. A., Steadman, D. W. & Witt, C. C. (2016), 'Predictable evolution toward flightlessness in volant island birds', *Proceedings of the National Academy of Sciences USA*, p.201522931.

11. Garcia-R., J. C. & Trewick, S. A. (2014), 'Dispersal and speciation in purple swamphens (Rallidae: *Porphyrio*)', *The Auk*, 132, 140–55.

12. Migrant birds would have occasionally brought continental diseases to isolated islands, but the lack of insect vectors meant that many islands remained disease-free until humans inadvertently imported mosquitos.

13. For example, New Zealand's strutting weka cannot fly, but it is sufficiently sprightly, pugnacious and fast-breeding that it has continued to survive in some parts of the New Zealand mainland.

14. Including Australia as a continent.

15. New Zealand is a bit different, because it is a fragment of continental land, and some New Zealand forms might have survived until the land eventually became reconnected to other continents.

16. Bacon, C. D. et al. (2015), 'Biological evidence supports an early and complex emergence of the Isthmus of Panama', *Proceedings of the National Academy of Sciences USA*, 112, 6110–15.

17. Leigh, E. G., O'Dea, A. & Vermeij, G. J. (2014), 'Historical biogeography of the Isthmus of Panama', *Biological Reviews*, 89, 148–72.

18. Bacon, C. D. et al. (2015), 'Biological evidence supports an early and complex emergence of the Isthmus of Panama', *Proceedings of the National Academy of Sciences USA*, 112, 6110–15.
19. The King Island, Kangaroo Island and Tasmanian emus did become extinct, but the Australian mainland species survived. The other very large Australian flightless birds were in the family Dromornithidae, which did become extinct when humans arrived.
20. Presumably, moas were slow, adopted ineffective defence strategies (e.g. those that might work against giant New Zealand eagles), did not recognize the danger, or their chicks were killed by predators that accompanied Maori colonists. Loss of giant insects after the arrival of kiore rats may also have deprived the chicks of food.
21. Relatively speaking; for a particular type of climate and habitat. Note that the past area (which may have changed since the last ice age) can be as relevant as the current area of a habitat.
22. Tilman, D. (2011), 'Diversification, biotic interchange, and the universal trade-off hypothesis', *American Naturalist*, 178, 355–71.

CHAPTER 7: EVOLUTION NEVER GIVES UP

1. Singer, M. C., Thomas, C. D. & Parmesan, C. (1993), 'Rapid human-induced evolution of insect–host associations', *Nature*, 366, 681–3.
2. The blue-eyed Marys are in the plant family Scrophulariaceae, and lousewort in the related plant family Orobanchaceae, hence they share some plant chemicals that make the switch possible. Evolution operates within a series of constraints, so they would not have been able to evolve to lay eggs on, and larvae to eat, lupins or pine trees in such a short space of time.
3. There was a genetic basis for the change; Singer, M. C., Ng, D. & Thomas, C. D. (1988), 'Heritability of oviposition preference and its relationship to offspring performance within a single insect population', *Evolution*, 42, 977–85.
4. Thomas C. D. (2005), 'Recent evolutionary effects of climate change', in *Climate Change and Biodiversity*, Lovejoy, T. E. & Hannah, L. (eds.), New Haven, CT: Yale University Press, pp. 75–88.
5. When I refer to 'bursts' and 'spurts' of evolution, I am referring to occasions when the rate of genotypic (the genes in each individual) change is relatively fast and results in rapid and substantial phenotypic change (in appearance, behaviour or physiology), which allows a population either to survive when conditions change, or to exploit new opportunities.

6. Severns, P. M. & Warren, A. D. (2008), 'Selectively eliminating and conserving exotic plants to save an endangered butterfly from local extinction', *Animal Conservation*, 11, 476–83; Severns, P. M. & Breed, G. A. (2014), 'Behavioral consequences of exotic host plant adoption and the differing roles of male harassment on female movement in two checkerspot butterflies', *Behavioral Ecology & Sociobiology*, 68, 805–14.

7. Bowers, M. D. & Richardson, L. L. (2013), 'Use of two oviposition plants in populations of *Euphydryas phaeton* Drury (Nymphalidae)', *Journal of the Lepidopterists' Society*, 67, 299–300.

8. Graves, S. D. & Shapiro, A. M. (2003), 'Exotics as host plants of the California butterfly fauna', *Biological Conservation*, 110, 413–33.

9. Singer, M. C., Thomas, C. D. & Parmesan, C. (1993), 'Rapid human-induced evolution of insect–host associations', *Nature*, 366, 681–3.

10. Bourn, N. A. D. & Thomas, J. A. (1993), 'The ecology and conservation of the brown argus butterfly *Aricia agestis* in Britain', *Biological Conservation*, 63, 67–74.

11. Pateman, R. M. et al. (2012), 'Temperature-dependent alterations in host use drive rapid range expansion in a butterfly', *Science*, 336, 1028–30.

12. Thomas, C. D. et al. (2001), 'Ecological and evolutionary processes at expanding range margins' *Nature*, 411, 577–81; Buckley, J., Butlin, R. K. & Bridle, J. R. (2012), 'Evidence for evolutionary change associated with the recent range expansion of the British butterfly, *Aricia agestis*, in response to climate change', *Molecular Ecology*, 21, 267–80; Bridle, J. R. et al. (2014), 'Evolution on the move: Specialization on widespread resources associated with rapid range expansion in response to climate change', *Proceedings of the Royal Society of London B: Biological Sciences*, 281, 20131800; Buckley, J. & Bridle, J. R. (2014), 'Loss of adaptive variation during evolutionary responses to climate change', *Ecology Letters*, 17, 1316–25.

13. As of December 2016, the international Paris climate change agreement aims to keep the average global temperature below 2°C (and preferably closer to 1.5°C) above pre-industrial conditions. However, the individual country pledges made to date are expected to lead to warming of approximately 2.7°C, if they are all fulfilled.

14. This is akin to the geneticist's concept of a 'ring species', where there is a geographic continuum of populations, each of which can reproduce with its neighbours, but if the extreme ends of the continuum meet up, they operate as though they are separate species.

15. Phillipps, Q. & Phillipps, K. (2014), *'Phillipps' Field Guide to the Birds of Borneo*, Oxford: John Beaufoy Publishing (3rd edition).

16. Laland, K. N., Odling-Smee, F. J. & Myles, S. (2010), 'How culture shaped the human genome: Bringing genetics and the human sciences together', *Nature Reviews Genetics*, 11, 137–48.

17. Hooke, R. L., Martín-Duque, J. F. & Pedraza, J. (2012), 'Land transformation by humans: A review', *GSA Today*, 22, 4–10.

18. Carroll, S. P. et al. (2007), 'Evolution on ecological time-scales, *Functional Ecology*, 21, 387–93.

19. Darimont, C. T. et al. (2009), 'Human predators outpace other agents of trait change in the wild', *Proceedings of the National Academy of Sciences, USA*, 106, 952–4.

20. Hemingway, J. & Ranson, H. (2000), 'Insecticide resistance in insect vectors of human disease', *Annual Review of Entomology*, 45, 371–91.

21. Roush, R. & Tabashnik, B. E. (eds.) (2012), *Pesticide Resistance in Arthropods*, Springer Science & Business Media.

22. Heap, I. (2014), 'Herbicide resistant weeds', in *Integrated Pest Management*, pp. 281–301, Netherlands: Springer.

23. Pelz, H. J. et al. (2005), 'The genetic basis of resistance to anticoagulants in rodents' *Genetics*, 170, 1839–47.

24. Berthold, P. et al. (1992), 'Rapid microevolution of migratory behaviour in a wild bird species', *Nature*, 360, 668–70; Karell, P. et al. (2011), 'Climate change drives microevolution in a wild bird', *Nature Communications*, 2, 208.

25. Hill, J. K., Griffiths, H. M. & Thomas, C. D. (2011), 'Climate change and evolutionary adaptations at species' range margins', *Annual Review of Entomology*, 56, 143–59.

26. Phillips, B. L. et al. (2006), 'Invasion and the evolution of speed in toads', *Nature*, 439, 803.

27. Saccheri, I. J. et al. (2008), 'Selection and gene flow on a diminishing cline of melanic peppered moths', *Proceedings of the National Academy of Sciences, USA*, 105, 16212–17.

28. Antonovics, J., Bradshaw, A. D. & Turner, R. G. (1971), 'Heavy metal tolerance in plants', *Advances in Ecological Research* 7, 1–85.

CHAPTER 8: THE PANGEAN ARCHIPELAGO

1. Arbogast, B. S. et al. (2006), 'The origin and diversification of Galapagos mockingbirds', *Evolution* 60, 370–82; Nietlisbach, P. et al. (2013),

'Hybrid ancestry of an island subspecies of Galápagos mockingbird explains discordant gene trees', *Molecular Phylogenetics and Evolution*, 69, 581–92.

2. Durham, W. H. (2012), 'What Darwin found convincing in Galápagos', *The Role of Science for Conservation*, 34, 1.

3. As new islands are continuously emerging as a result of volcanic activity and then erode away, the first colonists may actually have arrived on Galapagos islands that are now submerged.

4. Some further exchange of individuals would continue (gene flow), but immigration of new individuals from another island would be very small, relative to the size of the resident population on each island. Thus, the populations on each island could continue to diverge – although there would be potential for genes that are beneficial on all islands to spread between them.

5. Grant, P. R. & Grant, B. R. (2011), *How and Why Species Multiply: The Radiation of Darwin's Finches*, Princeton, New Jersey: Princeton University Press.

6. Lamichhaney, S. et al. (2015), 'Evolution of Darwin's finches and their beaks revealed by genome sequencing', *Nature*, 518, 371–5.

7. Lerner, H. R. et al. (2011), 'Multilocus resolution of phylogeny and timescale in the extant adaptive radiation of Hawaiian honeycreepers', *Current Biology*, 21, 1838–44.

8. The process of evolution into separate species in isolated geographic locations is known as 'allopatric speciation'.

9. Including New Zealand, but excluding New Guinea, which already supported terrestrial mammals. Once a species is flightless, it may lose the ability to colonize new islands (though not completely: volcanic activity, plate tectonics, coral growth and sea-level changes may connect and then disconnect islands).

10. Tilman, D. (2011), 'Diversification, biotic interchange, and the universal trade-off hypothesis', *American Naturalist*, 178, 355–71.

11. This excludes South American deer species that also live in Central and North America. The taxonomy of South American deer is still debated, so the final 'agreed' number of species may be slightly higher or lower than fourteen.

12. In the rodent subfamily Sigmodontinae.

13. Hughes, C. & Eastwood, R. (2006), 'Island radiation on a continental scale: exceptional rates of plant diversification after uplift of the Andes', *Proceedings of the National Academy of Sciences USA*, 103, 10334–9; Nevado, B. et al. (2016), 'Widespread adaptive evolution

during repeated evolutionary radiations in New World lupins', *Nature Communications*, 7, 12384.

14. Tilman, D. (2011), 'Diversification, biotic interchange, and the universal trade-off hypothesis', *American Naturalist*, 178, 355–71.

15. Zuk, M., Simmons, L. W. & Cupp, L. (1993), 'Calling characteristics of parasitized and unparasitized populations of the field cricket *Teleogryllus oceanicus*', *Behavioral Ecology and Sociobiology*, 33, 339–43.

16. Zuk, M., Rotenberry, J. T. & Tinghitella, R. M. (2006), 'Silent night: Adaptive disappearance of a sexual signal in a parasitized population of field crickets', *Biology Letters*, 2, 521–4.

17. Johnston, R. F. & Selander, R. K. (1964), 'House sparrows: Rapid evolution of races in North America', *Science*, 144, 548–50; Schrey, A. W. et al. (2011), 'Broad-scale latitudinal patterns of genetic diversity among native European and introduced house sparrow (*Passer domesticus*) populations', *Molecular Ecology*, 20, 1133–43; Liebl, A. L. et al. (2015), 'Invasion genetics: Lessons from a ubiquitous bird, the house sparrow *Passer domesticus*', *Current Zoology*, 61, 465–76; Huey, R. B. et al. (2000), 'Rapid evolution of a geographic cline in size in an introduced fly', *Science*, 287, 308–9; Balanyá, J. et al. (2006), 'Global genetic change tracks global climate warming in *Drosophila subobscura*', *Science*, 313, 1773–5.

18. Langkilde, T. (2009), 'Invasive fire ants alter behavior and morphology of native lizards', *Ecology*, 90, 208–17.

19. Some seeds may survive for several years before they germinate, so the total number of generations may be less than the number of years.

20. Montesinos, D., Santiago, G. & Callaway, R. M. (2012), 'Neo-allopatry and rapid reproductive isolation', *American Naturalist*, 180, 529–33.

21. Feder J. L. (1998), in Howard D. J. & Berlocher S. H. (eds.), *Endless Forms: Species and Speciation*, New York: Oxford University Press; Feder, J. L. et al. (2003), 'Evidence for inversion polymorphism related to sympatric host race formation in the apple maggot fly, *Rhagoletis pomonella*', *Genetics*, 163, 939–53.

22. The first apple orchard in North America was planted in 1625, in Boston. However, the evolution of the apple fly is not likely to have commenced until the continent-wide proliferation of apples during the nineteenth century.

23. Forbes, A. A. et al. (2009), 'Sequential sympatric speciation across trophic levels', *Science*, 323, 776–9; Hood, G. R. et al. (2015), 'Sequential divergence and the multiplicative origin of community diversity',

Proceedings of the National Academy of Sciences USA, 112, E5980–E5989.

24. It is now accepted that one species can sometimes turn into two in the absence of geographic separation of the populations, a process known as sympatric speciation. The frequency of these events is still debated, as well as whether apple flies are quite 'there' yet. Berlocher, S. H. & Feder, J. L. (2002), 'Sympatric speciation in phytophagous insects: Moving beyond controversy?', *Annual Review of Entomology*, 47, 773–815.

25. Geneticists do not all agree on the definition of a species: some prefer a definition in which there is (almost) no possibility of successful reproduction with other species (the apple fly has not yet reached this stage); others consider separate species to be 'types' that remain distinct, despite some ongoing mating and gene flow between the two (the apple fly has apparently reached this stage).

CHAPTER 9: HYBRID

1. Harris, S. A. (2002), 'Introduction of Oxford ragwort, *Senecio squalidus* L. (Asteraceae), to the United Kingdom', *Watsonia*, 24, 31–43.

2. Not everyone would count it as a 'full species', but it is a self-perpetuating genetic form that no longer interbreeds with its parental species through geographic isolation; Abbott, R. J. et al. (2003), 'Plant introductions, hybridization and gene flow', *Philosophical Transactions of the Royal Society B*, 358, 1123–32; Abbott, R. J. et al. (2009), 'Recent hybrid origin and invasion of the British Isles by a self-incompatible species, Oxford ragwort (*Senecio squalidus* L., Asteraceae)', *Biological Invasions*, 11, 1145–58.

3. Lowe, A. J. & Abbott, R. J. (2004), 'Reproductive isolation of a new hybrid species, *Senecio eboracensis* Abbott & Lowe (Asteraceae)', *Heredity*, 92, 386–95; Abbott, R. J. & Lowe, A. J. (2004), 'Origins, establishment and evolution of new polyploid species: *Senecio cambrensis* and *S. eboracensis* in the British Isles', *Biological Journal of the Linnean Society*, 82, 467–74.

4. Abbott, R. J. & Lowe, A. J. (2004), 'Origins, establishment and evolution of new polyploid species: *Senecio cambrensis* and *S. eboracensis* in the British Isles', *Biological Journal of the Linnean Society*, 82, 467–74; Hegarty, M. J., Abbott, R. J. & Hiscock, S. J. (2012), 'Allopolyploid speciation in action: The origins and evolution of *Senecio cambrensis*', in Soltis, P. S. & Soltis, D. E. (eds.), *Polyploidy and Genome Evolution*, pp. 245–70, Heidelberg: Springer.

5. Ainouche, M. & Gray, A. (2016), 'Invasive *Spartina*: Lessons and challenges', *Biological Invasions*, 18, 2119–22.

6. Thomas, C. D. (2015), 'Rapid acceleration of plant speciation during the Anthropocene', *Trends in Ecology & Evolution*, 30, 448–55.

7. This calculation is based on multiplying up the world land surface, excluding ice and deserts. This number should be regarded as an order-of-magnitude estimate.

8. Many of these hybrids are between so-called 'native' species, but many 'native' British plants live where they do only because humans have altered the landscape and transported species for millennia.

9. Milne, R. I. & Abbott, R. J. (2000), 'Origin and evolution of invasive naturalized material of *Rhododendron ponticum* L. in the British Isles', *Molecular Ecology*, 9, 541–56.

10. Matsuoka, Y. (2011), 'Evolution of polyploid *Triticum* wheats under cultivation: The role of domestication, natural hybridization and allopolyploid speciation in their diversification', *Plant & Cell Physiology*, 52, 750–64.

11. Seijo, G. et al. (2007), 'Genomic relationships between the cultivated peanut (*Arachis hypogaea*, Leguminosae) and its close relatives revealed by double GISH', *American Journal of Botany*, 94, 1963–71.

12. Schmidt, R. & Bancroft, I. (eds.) (2011), *Genetics and Genomics of the Brassicaceae*, New York: Springer-Verlag.

13. Soltis, P. S. & Soltis, D. E. (2009), 'The role of hybridization in plant speciation', *Annual Review of Plant Biology*, 60, 561–88.

14. Morgan-Richards, M. et al. (2004), 'Interspecific hybridization among *Hieracium* species in New Zealand: Evidence from flow cytometry', *Heredity*, 93, 34–42.

15. Sankararaman, S. et al. (2014), 'The genomic landscape of Neanderthal ancestry in present-day humans', *Nature*, 507, 354–7.

16. Huerta-Sánchez, E. et al. (2014), 'Altitude adaptation in Tibetans caused by introgression of Denisovan-like DNA', *Nature*, 512, 194–7; Harrison, R. G. & Larson, E. L. (2014), 'Hybridization, introgression, and the nature of species boundaries', *Journal of Heredity*, 105 (S1), 795–809.

17. Mavárez, J. et al. (2006), 'Speciation by hybridization in *Heliconius* butterflies', *Nature*, 441, 868–71; Arias, C. F. et al. (2014), 'Phylogeography of *Heliconius cydno* and its closest relatives: Disentangling their origin and diversification', *Molecular Ecology*, 23, 4137–52; Heliconius Genome Consortium (2012), 'Butterfly genome reveals promiscuous exchange of mimicry adaptations among species', *Nature*, 487, 94–8;

Kozak, K. M. et al. (2015), 'Multilocus species trees show the recent adaptive radiation of the mimetic *Heliconius* butterflies', *Systematic Biology*, syv007.

18. Pollinger, J. P. et al. (2011), 'A genome-wide perspective on the evolutionary history of enigmatic wolf-like canids', *Genome Research*, 21, 1294–305.

19. Scriber, J. M., & Ording, G. J. (2005), 'Ecological speciation without host plant specialization: Possible origins of a recently described cryptic *Papilio* species', *Entomologia Experimentalis et Applicata*, 115, 247–63; Cong, Q. et al. (2015), 'Tiger swallowtail genome reveals mechanisms for speciation and caterpillar chemical defense', *Cell Reports*, 16, 910–19.

20. Amaral, A. R. et al. (2014), 'Hybrid speciation in a marine mammal: The Clymene dolphin (*Stenella clymene*)', *PLOS ONE*, 9, e83645.

21. Miller, W. et al. (2012), 'Polar and brown bear genomes reveal ancient admixture and demographic footprints of past climate change', *Proceedings of the National Academy of Sciences USA*, 109, E2382–E2390; Cahill, J. A. et al. (2015), 'Genomic evidence of geographically widespread effect of gene flow from polar bears into brown bears', *Molecular Ecology*, 24, 1205–17.

22. Gogarten, J. P., Doolittle, W. F. & Lawrence, J. G. (2002), 'Prokaryotic evolution in light of gene transfer', *Molecular Biology and Evolution*, 19, 2226–38.

23. Schwarz, D. et al. (2007), 'A novel preference for an invasive plant as a mechanism for animal hybrid speciation', *Evolution*, 61, 245–56.

24. Schwarz, D. et al. (2005), 'Host shift to an invasive plant triggers rapid animal hybrid speciation', *Nature*, 436, 546–9.

25. Senn, H. V. & Pemberton, J. M. (2009), 'Variable extent of hybridization between invasive sika (*Cervus nippon*) and native red deer (*C. elaphus*) in a small geographical area', *Molecular Ecology*, 18, 862–76.

26. http://www.nonnativespecies.org/factsheet/downloadFactsheet.cfm?speciesId=725.

27. Hedrick, P. W. (2009), 'Conservation genetics and North American bison (*Bison bison*)', *Journal of Heredity*, 100, 411–20.

28. Fonseca, F. (2016), 'Grand Canyon weighs killing, capturing bison to cut numbers', *Kansas City Star*, 26 February 2016.

29. Plumb, B. et al. (2016), 'Grand Canyon bison nativity, genetics, and ecology: Looking forward', Natural Resource Report NPS/NRSS/BRD/NRR—2016/1226, Fort Collins, Colorado: US Department of the Interior National Park Service.

30. Soubrier, J. et al. (2016), 'Early cave art and ancient DNA record the origin of European bison', *Nature Communications*, 7, 13158.
31. Microbes can be transported as dust-like particles in the air, so a higher proportion of them have nearly global distributions. This would probably always have been the case.
32. Thomas, C. D. (2015), 'Rapid acceleration of plant speciation during the Anthropocene', *Trends in Ecology & Evolution*, 30, 448–55.
33. Wikipedia provides five additional suggestions for extinct North American higher plants which deserve formal IUCN assessment of their taxonomic status, whether any survive in the wild, and whether any individuals or seeds are available in botanic gardens, seed banks, etc; https://en.wikipedia.org/wiki/List_of_extinct_plants#Americas (accessed 14 October 2016).

CHAPTER 10: THE NEW NATURAL

1. Accepting that some world-views (e.g., animism, Buddhism) take a somewhat more integrated perspective.
2. Darwin, C. & Wallace, A. (1858), 'On the tendency of species to form varieties; and on the perpetuation of varieties and species by natural means of selection', *Journal of the Proceedings of the Linnean Society of London, Zoology*, 3, 45–62; Darwin, C. (1859), *On the Origin of Species by Means of Natural Selection, or the preservation of favoured races in the struggle for life*, London: John Murray; Darwin, C. (1871), *The Descent of Man, and selection in relation to sex*, London: John Murray.
3. Others represent humans living in rose-tinted harmony with nature, usually at some time in the past or far away, serving to emphasize that humans and nature used to be at one, but now we are at loggerheads. This is equally fallacious.
4. Roberts, C. (2007), *The Unnatural History of the Sea*, London: Gaia/ Octopus Publishing; Kolbert, E. (2014), *The Sixth Extinction: An Unnatural History*, London: Bloomsbury Publishing. Despite my disapproval of 'unnatural' in the titles of these two books, they are both wonderfully written and informative accounts of biological losses and changes in the human era.
5. Oxygen depletion normally involves feedback between geological and biological processes.
6. Because of continued mating between closely related species, some of our genes may have separated considerably more than 7 million years

ago, and some more recently than 6 million. Wakeley, J. (2008), 'Complex speciation of humans and chimpanzees', *Nature*, 452, E3–E4; Langergraber, K. E. et al. (2012), 'Generation times in wild chimpanzees and gorillas suggest earlier divergence times in great ape and human evolution', *Proceedings of the National Academy of Sciences USA*, 109, 15716–21; Arnold, M. L. et al. (2015), 'Divergence-with-gene-flow – what humans and other mammals got up to', in *Reticulate Evolution* (pp. 255–95), Switzerland: Springer International Publishing.

7. Almécija, S., Smaers, J. B. & Jungers, W. L. (2015), 'The evolution of human and ape hand proportions,' *Nature Communications*, 6, 7717.

8. Love and protection is not engendered by direct genetic recognition, because adopted children also receive love and attention. Similarly, children and other relatives are not loved less if they do not themselves reproduce. Nonetheless, the molecular bases of loving, protecting and provisioning behaviours can have evolved only if, on average, they increase the evolutionary success of individuals bearing those inclinations. Offspring also love parents, developing strong bonds that ensure their own survival and eventual reproduction.

9. Alberti, M. et al. (2017), 'Global urban signatures of phenotypic change in animal and plant populations', *Proceedings of the National Academy of Sciences USA*, 201606034.

10. Ecologists disagree whether there are residual negative impacts of radiation (as opposed to effects of land abandonment) on animal populations in the most radioactive areas of the Chernobyl region but agree that numbers are high in the lower-radiation parts of the exclusion zone; Møller, A. P. & Mousseau, T. A. (2013), 'Assessing effects of radiation on abundance of mammals and predator–prey interactions in Chernobyl using tracks in the snow', *Ecological Indicators*, 26, 112–16; Deryabina, T. G. et al. (2015), 'Long-term census data reveal abundant wildlife populations at Chernobyl', *Current Biology*, 25, R824–R826.

11. If the forest is allowed to develop for several centuries, introduced trees like locust and box elder (which are successfully colonizing previously open areas) are likely to be rarer than at present but to remain part of the tree flora.

12. Werdelin, L. (2013), 'King of Beasts', *Scientific American*, 309, 34–9.

13. Roberts, R. G. et al. (2001), 'New ages for the last Australian megafauna: Continent-wide extinction about 46,000 years ago', *Science*, 292, 1888–92.

14. Many oceanic islands were colonized more recently, but they constitute little of the Earth's total land area (Madagascar approximately 0.4 per

cent, New Zealand approximately 0.2 per cent of the land surface). Impacts in the world's oceans are generally more recent, but also irreversible, given the numbers of species that have been moved from one ocean to another.

15. Excluding microbial communities inside the Earth's crust, beneath the Antarctic ice sheet, or in similar places that are more or less sealed away from human influence.

16. For a hard-hitting and amusing account, I recommend Thompson, K. (2014), *Where Do Camels Belong? The Story and Science of Invasive Species*, London: Profile Books.

17. When King Canute forbade the tide from coming up the beach, he was allegedly demonstrating that there were events that kings could not prevent, rather than the more popular version in which he is represented as stupidly imagining that he might be able to stop the tide.

CHAPTER II: NOAH'S EARTH

1. Villablanca, F. (2010), *Monarch Alert Annual Report: Overwintering Population 2009–2010*, Cal Poly State University.

2. Millar, C. I. (1998), 'Reconsidering the conservation of Monterey pine', *Fremontia*, 26 (3), 12–16. I refer colloquially to glacial maximum conditions as 'ice ages', even though the entire Pleistocene epoch of alternating colder and warmer periods can be thought of as one extended ice age.

3. Berg, P., 'Radiata pine', *Te Ara – The Encyclopedia of New Zealand*; http://www.TeAra.govt.nz/en/radiata-pine (2012 update).

4. Weiss, S. B. (2011), *Management Plan for Monarch Grove Sanctuary: Site Assessment and Initial Recommendations*, Menlo Park, CA: Creekside Center for Earth Observation.

5. Less the blue gum, which has a larger original distribution than the Monterey pine and is also planted commercially in Australia.

6. This could threaten some species in the South African fynbos, parts of which are being changed by introduced trees. In such a situation, controlling invading trees may be desirable (to save species, rather than to keep the vegetation unaltered) as a holding plan, while working on a long-term solution, such as biocontrol (releasing insects, fungi and pathogens) to reduce the vigour or reproduction of the invading trees.

7. Smith, S. E. et al. (2013), 'The past, present and potential future distributions of cold-adapted bird species', *Diversity & Distributions*, 19, 352–62.

8. Bennett, K. D., Tzedakis, P. C. & Willis, K. J. (1991), 'Quaternary refugia of north European trees', *Journal of Biogeography*, 18, 103–115; Tzedakis, P. C. et al. (2002), 'Buffered tree population changes in a Quaternary refugium: Evolutionary implications', *Science*, 297, 2044–7.

9. Early, R. & Sax, D. F. (2014), 'Climatic niche shifts between species' native and naturalized ranges raise concern for ecological forecasts during invasions and climate change', *Global Ecology & Biogeography*, 23, 1356–65.

10. Roberts, C. (2007), *The Unnatural History of the Sea*, London: Gaia/Octopus Publishing.

11. This applies particularly to lesser short-tailed bats, given that there may be a great deal to learn (of potential use to humans) about their take-off mechanics and mode of squirming-crawling movement across the ground.

12. Biodiversity refers to genetic, ecosystem (habitat) and species levels of diversity. Since no one has much information about the genetic diversity of most wild species, this usually ends up being an account of the species and habitats in each country.

13. I am referring to 'genes' in a colloquial sense, meaning unique alleles, or any other form of genetic variation.

14. Gibson, L. G. & Yong, D. L. (2017), 'Saving two birds with one stone: Solving the quandary of introduced, threatened species', *Frontiers in Ecology and the Environment*, DOI: 10.1002/fee.1449.

15. Donlan, C. J. et al. (2006), 'Pleistocene rewilding: An optimistic agenda for twenty-first-century conservation', *American Naturalist*, 168, 660–81.

EPILOGUE: ONE MILLION YEARS AD

1. Ocean acidification and the possibility that deoxygenation might become widespread could generate a different outcome in the oceans.

2. The rate of bird extinction since the year 1900 has been 132 species lost per 'million species years'; Pimm, S. L. et al. (2014), 'The biodiversity of species and their rates of extinction, distribution, and protection', *Science*, 344, p.1246752. This is the number of species that would go extinct in a year if there were a million species (or the number that would go extinct if we observed a thousand species for a thousand years). If the existing approximately 10,500 bird species show this level of extinction for the next thousand years, then we can expect approximately a further 12 per cent of species to become extinct. Pimm et al.

quote a hundred species becoming extinct per 'million species years' as an approximate estimate of documented extinction across several taxonomic groups.

3. The IUCN Red List of Threatened Species summary statistics; http://www.iucnredlist.org/about/summary-statistics#How_many_threatened, accessed 1 January 2017.

4. Local diversity (per square metre) has declined in fields with intensive agriculture and where the land is covered in concrete. This is desirable because we must live somewhere and it is essential to feed the world population, which is done most efficiently in relatively weed- and pest-free crops. However, it is the number of species per region that is likely to drive evolutionary diversification. On this scale, diversity is increasing.

5. Moyle, R. G. et al. (2009), 'Explosive Pleistocene diversification and hemispheric expansion of a "great speciator"', *Proceedings of the National Academy of Sciences USA*, 106, 1863–8.

Index

stencil

GREEN THE CITY

UNDER C.A.N.

mc c

1 _____

2 _____
_____
_____

comparies in ~ len
Secondary Schools.
+ private schools MHSG
Foundations.

Aston   Universities
award Nottm West.
& Nottingham +

middleham &
won to past
            present

live an open
rather than
            curted

Considering phd. for future
since MA explored & drawn

length of lessons?

link to spewen of
life on earth
not fixed

life in spewenun

He not busy being
born is busy dying!

Regulatory ot have?

wage?

length ot contract?

location?

trainer or teacher?

specific curriculum or
open to discussion?

existing materials?

resources - library access
etc?